Springer

Progress in the Chemistry
of Organic Natural Products

Founded by L. Zechmeister

96

Progress in the Chemistry of Organic Natural Products

Asymmetric Organocatalysis in Natural Product Syntheses

Author:
Mario Waser

Springer

Prof. A. Douglas Kinghorn, College of Pharmacy,
Ohio State University, Columbus, OH, USA

em. Univ.-Prof. Dr. H. Falk, Institut für Organische Chemie,
Johannes-Kepler-Universität, Linz, Austria

Prof. Dr. J. Kobayashi, Graduate School of Pharmaceutical Sciences,
Hokkaido University, Sapporo, Japan

ISSN 2191-7043 ISSN 2192-4309 (electronic)
ISBN 978-3-7091-1162-8 ISBN 978-3-7091-1163-5 (eBook)
DOI 10.1007/978-3-7091-1163-5
Springer Wien Heidelberg New York Dordrecht London

Library of Congress Control Number: 2012938965

Printed on acid-free paper

Springer is part of Springer Science+Business Media (www.springer.com)

Acknowledgments

I am very grateful to Prof. Dr. Heinz Falk, not only for all his support during the preparation of this manuscript, but also for being a wonderful mentor during all stages of my career. I am also very thankful to Prof. Dr. Norbert Müller for all his very kind and helpful support over the last few years. Additional thanks go to all the members of the Institute of Organic Chemistry in Linz, especially to my very talented and gifted PhD students Katharina Gratzer and Richard Herchl. Our own work has been supported generously by the Austrian Science Funds (FWF) for several years (Project No. P22508-N17). Last but not least, I would like to express much gratitude to my girlfriend Dr. Manuela Haunschmidt for tolerating and supporting my long working hours in the laboratory and in front of the computer.

Contents

Listed in PubMed

Contributor

Mario Waser Institute of Organic Chemistry, Johannes Kepler University Linz, Altenbergerstrasse 69, 4040 Linz, Austria

About the Author

Mario Waser was born in Steyr, Austria in 1977. After studying chemistry at the *Johannes Kepler* University (JKU) Linz, Austria, he finished his Ph.D. thesis in 2005 in the group of Prof. Dr. *Heinz Falk*, working on the synthesis of second-generation hypericin-based photosensitizers. After a postdoctoral stay in the group of Prof. Dr. *Alois Fürstner* at the *Max-Planck* Institut für Kohlenforschung (Mülheim, Germany), investigating the first total syntheses of iejimalide B and iejimalide A, he spent two years as a research and development chemist for DSM Linz. Since the summer of 2009 he has held the position of Assistant Professor at JKU Linz, where he is currently working on his habilitation. Dr. *Waser's* main research interests are focused in the field of organic synthesis chemistry with a special emphasis on ammonium ylide-mediated (dia)-stereoselective reactions and the design of new tartaric acid-derived organocatalysts.

1 Introduction

Over the ages, organisms have developed the capacity to elaborate a fascinating variety of natural products with an almost infinite diversity in structure and biological activity. It is impressive (and sometimes may also be a bit frustrating) to synthesis-oriented organic chemists to recognize the ease with which Nature biosynthesizes such important compounds like nucleic acids, saccharides, amino acids (proteins), or various highly complex secondary metabolites (*1*).

Due to the high chemical diversity of compounds available from natural sources, the identification and isolation of novel biologically active natural products represents a major goal in contemporary biomedical and agrochemical science and a large percentage of today's major drugs have their origins in Nature. However, not all of the potentially useful natural compounds can be isolated readily or in such large quantities as can amino acids or saccharides. Accordingly, the paucity of such natural products may require a (total) synthesis approach to obtain sufficient quantities for initial biological investigations, and, if promising, for further development.

The field of natural product synthesis is definitely one of the most challenging and attractive areas of organic chemistry and numerous contributions focusing on the development of synthesis routes for natural products are reported constantly (*2–4*). Among the different types of transformations that are necessary to successfully achieve a complex total synthesis, those enabling the stereoselective introduction of a stereogenic center have attracted special interest.

The field of asymmetric synthesis has made spectacular progress over the last few decades. Among the various ways of creating enantiomerically enriched products, catalytic methods are considered to be the most appealing as the use of stoichiometric amounts of valuable chiral reagents can be avoided, thus resulting in highly efficient approaches. New methods have been emerging recently, enabling more selective, environmentally friendly, and economically more cost-effective transformations. Besides enzymatic and metal-catalyzed asymmetric transformations, the use of substoichiometric amounts of organic molecules (with so-called "organocatalysts") has proven to possess an enormous potential for the catalysis of stereoselective reactions (*5*, *6*).

M. Waser, *Asymmetric Organocatalysis in Natural Product Syntheses*,
Progress in the Chemistry of Organic Natural Products 96,
DOI 10.1007/978-3-7091-1163-5_1,

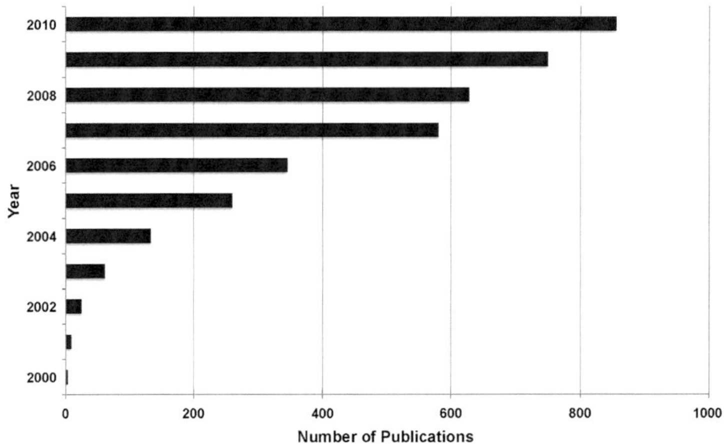

Fig. 1 Annual number of publications covering the topic organocatalysis (SciFinder®, Chemical Abstracts Service, Columbus, OH, U.S.A.)

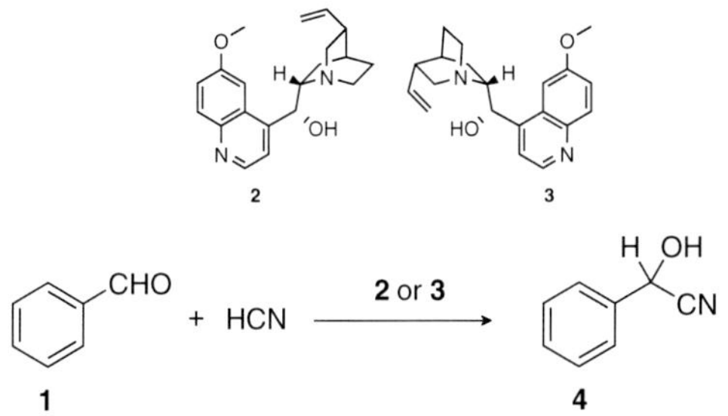

Scheme 1 *Cinchona* alkaloid-mediated addition of HCN to **1**

Although the use of small molecules for the catalysis of a variety of organic reactions has been known for decades (*7–26*), it was only just over 10 years ago, when the seminal publications of *MacMillan et al.* (*27*) and *List*, *Barbas*, and *Lerner* (*28*) really set the stage for a new trend in organic synthesis. It is also due to *David MacMillan* that the term "organocatalysis" currently has become a catchword for a whole field of research in (organic) chemistry (*27*). As illustrated in Fig. 1, the annual number of scientific publications covering the topic organocatalysis has increased significantly since these seminal reports in 2000.

At this point it should be mentioned that, although the two above-cited publications are often considered to represent the genesis of modern organocatalysis, several very important contributions appeared long before 2000, but have not been considered to refer to organocatalysis, because this new term did not exist at the time of their appearance (*7*, *12*, *14–16*, *19*). From a historical point of view, asymmetric organocatalysis can be dated back to the beginning of the last century when *Breding* performed the addition of HCN to benzaldehyde (**1**) in the presence of quinine (**2**) or quinidine (**3**) (Scheme 1) (*10*). Although the enantioselectivity of this reaction was less than 10%, it was one of the conceptually groundbreaking investigations in this field.

Later on, in the 1950s, *Prelog* reinvestigated this reaction (*15*) and in the late 1950s *Pracejus* reported one of the first highly enantioselective reactions ever, by adding methanol to methyl phenyl ketene (**5**) in the presence of *O*-acetylquinine (**6**) (Scheme 2) (*16*).

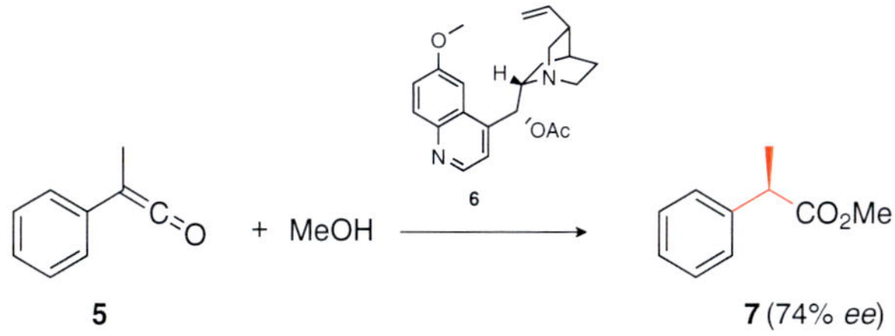

Scheme 2 *Cinchona* alkaloid-mediated addition of MeOH to **5**

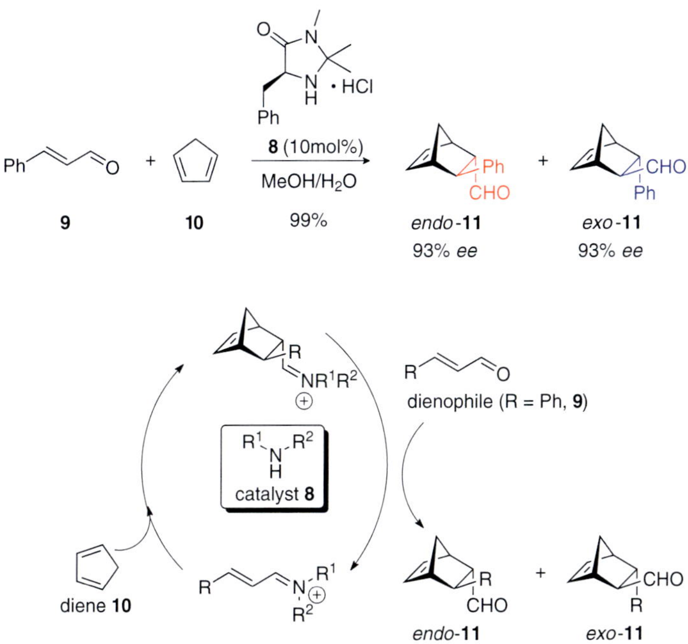

Scheme 3 Chiral imidazolium salt-catalyzed *Diels-Alder* reaction proceeding *via* an iminium mechanism (*27*)

In terms of modern organocatalysis, the publications of *MacMillan et al.* and *List et al.* in 2000 set the stage for two of the most important activation mechanisms employed in organocatalysis today: iminium catalysis (*27*) and enamine catalysis (*28*). While *MacMillan* and co-workers used the chiral imidazolium salt **8** to

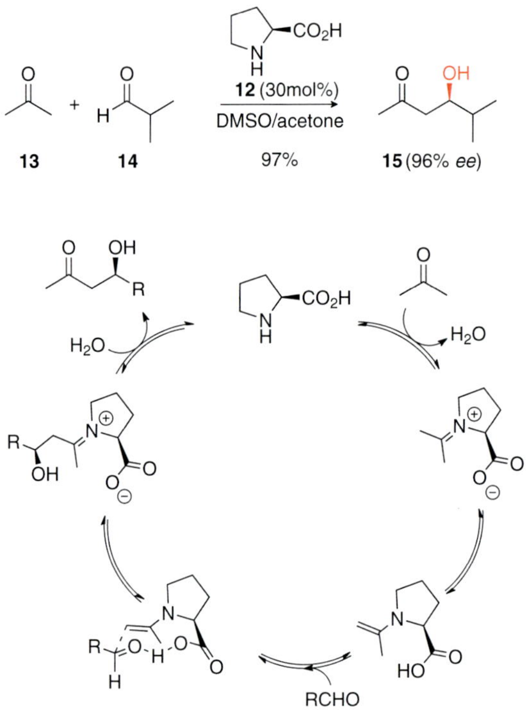

Scheme 4 Proline (**12**)-catalyzed aldol reaction proceeding *via* an enamine mechanism (*28*)

activate α,β-unsaturated aldehydes for asymmetric *Diels-Alder* reactions by a reversible formation of an iminium ion (Scheme 3), *List*, *Lerner*, and *Barbas* used the natural amino acid proline (**12**) for enantioselective cross-aldol reactions between acetone (**13**) and different aldehydes proceeding *via* an enamine mechanism (Scheme 4).

As shown in Scheme 4, proline (**12**) is considered to act as a bifunctional catalyst in these types of aldol reactions. On the one hand it activates the nucleophile *via* enamine formation, but, on the other hand, activation and coordination of the electrophile *via* the carboxylic acid leads to the formation of a defined transition state, explaining the observed high selectivity.

Although new activation modes and catalytic principles were first investigated using simple standard benchmark reactions, organocatalysis has very shortly thereafter shown also its high potential for the syntheses of biologically active complex natural products and natural product analogs, with numerous impressive examples for the application of organocatalytic transformations in this field having been reported already (*29–33*).

Therefore, it is the aim of this contribution to give the reader an overview of the most impressive applications of organocatalytic reactions in the synthesis of natural products. It is not intended to give a full encyclopedic coverage of the literature but to illustrate the high potential of organocatalysis in the field of natural product synthesis by giving selected examples. The main focus will be on the key organocatalytic steps for each (multi-step) synthesis described, whereas other often particularly innovative transformations will be omitted, as this would be beyond the scope of this volume.

This contribution will be divided according to activation mechanisms used to achieve the targeted transformation and the reaction type itself. However, some caution is necessary. As already shown in the case of the proline-catalyzed intermolecular aldol reaction (Scheme 4), **12** can be considered to act as a bifunctional catalyst. Therefore, a strict classification according to just one single activation mechanism will not always be possible and very often activation modes like *e.g.* enamine formation are accompanied with additional interactions, such as *e.g.* hydrogen bonding.

2 Enamine Catalysis

In 1971, *Eder*, *Sauer*, and *Wiechert* at Schering (*12*) and *Hajos* and *Parrish* at Hoffmann-La Roche (*13*, *14*) independently reported a proline-catalyzed intramolecular aldol reaction of the triketone **16** as the key step in the synthesis of the diketone **17**, a highly important intermediate in steroid synthesis. Remarkably, *Hajos* and *Parrish* obtained the diketone **18** in excellent yield and enantioselectivity with only 3 mol% of catalyst (Scheme 5). Acid-mediated dehydration then furnished the targeted **17**. The accepted transition state for this reaction is believed to include one proline molecule as elucidated by *List* and *Houk* (*21*, *34*).

It is worth noting that, although *Hajos* and *Parrish* considered this reaction to be "a simplified model of a biological system in which (*S*)-proline plays the role of an enzyme", which represented a unique and groundbreaking approach for the introduction of stereogenic centers, this methodology was not developed further for almost 30 years until *List*, *Barbas*, and *Lerner* published their breakthrough report on the intermolecular aldol reaction, as depicted in Scheme 4 (*28*).

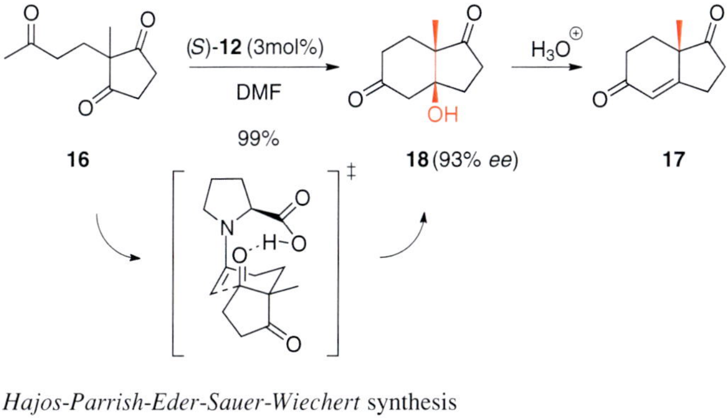

Scheme 5 *Hajos-Parrish-Eder-Sauer-Wiechert* synthesis

M. Waser, *Asymmetric Organocatalysis in Natural Product Syntheses*,
Progress in the Chemistry of Organic Natural Products 96,
DOI 10.1007/978-3-7091-1163-5_2, © Springer-Verlag Wien 2012

Presently, enamine catalysis, meaning the utilization of carbonyl groups by catalyzing their reactions with primary or secondary amines *via* enamine derivatives, plays a fundamental role in organic synthesis (*5, 6, 8, 9, 35, 36*). As enamine catalysis can be viewed as reducing the function and activation mode of aldolase enzymes to small organic molecules, it can be stated beyond doubt that this methodology represents one of the most powerful methods for the stereoselective α-functionalization of aldehydes and ketones currently known (*35, 36*). The following sections of this chapter will be subdivided into different transformation types that can be achieved by enamine catalysis.

2.1 Aldol Reactions

Aldol and *Mannich*-type reactions were the first systematically investigated applications for enamine catalysis. Aldol reactions belong to the most commonly applied C–C bond-forming reactions (*37, 38*) allowing for the construction of chiral building blocks for the syntheses of a variety of structurally complex molecules. These reactions are very often carried out using a preformed enolate (indirect aldol reaction) in combination with a chiral catalyst or using covalently bond chiral auxiliaries (*38–41*). Very often asymmetrically catalyzed reactions are carried out in the presence of metal catalysts or chiral *Lewis* bases (*41*). Besides the indirect approach, the direct aldol reaction between two unmodified carbonyl compounds is of great interest as it avoids the formation and handling of an enolate equivalent (*42, 43*). The seminal publication of *List*, *Barbas*, and *Lerner* for the proline (**12**)-catalyzed aldol reaction in 2000 (*28*) set the starting point for a number of impressive applications of enamine-type direct aldol reactions in natural product syntheses.

2.1.1 Ketone Donors in Intermolecular Aldol Reactions

Reactions between ketone donors and aldehyde acceptors strongly depend on the nature of the aldehyde. While α-disubstituted aldehydes normally react easily, unbranched ones often undergo self-addition reactions. *List et al.* reported one of the first examples of a direct aldol addition of ketones to α-unbranched aldehydes *en route* to a natural product in 2001 (*44*). The operationally simple reaction between **13** and **19** in the presence of catalytic amounts of (*S*)-**12** furnished the enantiomerically enriched β-hydroxy-ketone **20** in moderate yield. The reduced yield can be rationalized by the concomitant formation of the condensation product **21**, which is one of the limiting factors in such reactions (besides the self reaction of α-unbranched aldehydes). Intermediate **20** can then be further converted to the bark beetle pheromone (*S*)-ipsenol (**22**) in two more steps (Scheme 6).

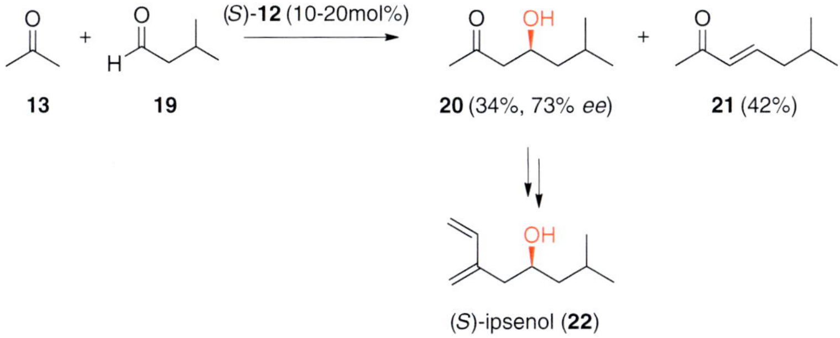

Scheme 6 (*S*)-Proline (**12**)-catalyzed formation of a key intermediate in the synthesis of (*S*)-ipsenol (**22**)

A methodologically similar approach was successfully utilized independently for the synthesis of the oviposition attractant pheromone of the female *Culex* mosquito (−)-(5*R*,6*S*)-6-acetoxyhexadecanolide (**28**), by *Kotsuki et al.* (*45*) and *Li et al.* (*46*). While the *Li* group carried out a direct aldol reaction between undecanal (**23**) and cyclopentanone (**24**) to obtain the desired isomer **25** in excellent enantio- and diastereoselectivity, the *Kotsuki* group introduced the stereogenic centers by a reaction between **24** and the dithiane **26** under solvent-free conditions (Scheme 7).

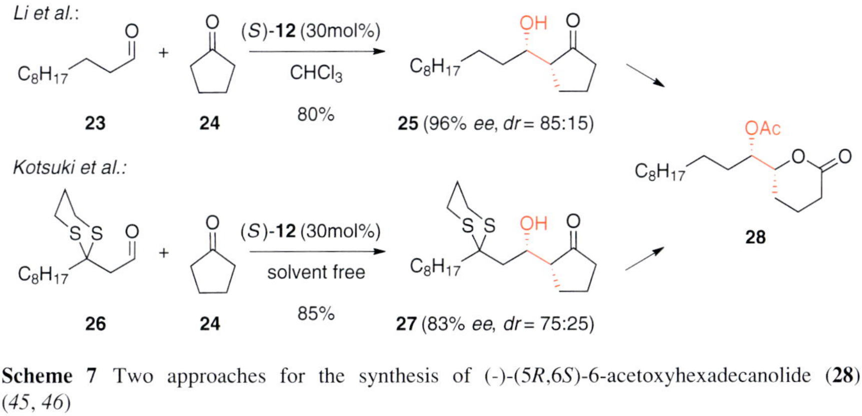

Scheme 7 Two approaches for the synthesis of (-)-(5*R*,6*S*)-6-acetoxyhexadecanolide (**28**) (*45*, *46*)

The amine-catalyzed aldol reaction between ketone donors and α-disubstituted aldehydes normally proceeds much more easily and with excellent enantioselectivity, which was demonstrated impressively in the synthesis of the southern part of the highly cytotoxic potential anticancer drug epothilone B (**29**) (*47*, *48*) by *Avery* and *Zheng* (*49*) (Scheme 8). In this case (*R*)-proline was the catalyst of choice to introduce the secondary alcohol group in high enantioselectivity early in the synthesis sequence.

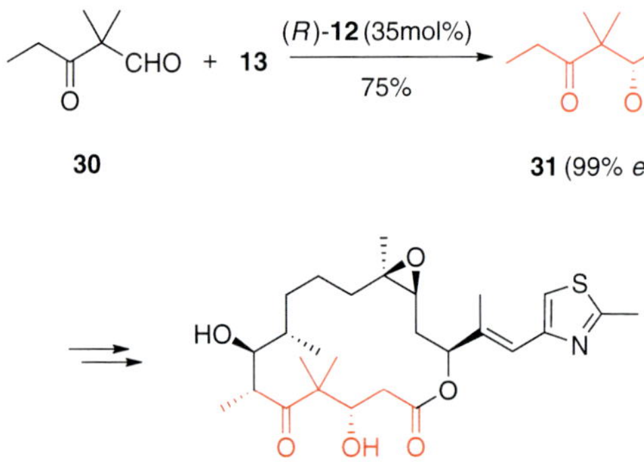

30

31 (99% *ee*)

Epothilone B (**29**)

Scheme 8 (*R*)-Proline (**12**)-catalyzed aldol reaction in the synthesis of the southern part of epothilone B (**29**)

This example demonstrates the strength and versatility of enamine-catalyzed aldol reactions between ketone donors and aldehydes for the synthesis of key natural product synthons in a very impressive way. Like other routinely used asymmetric organic transformations that are applied commonly in total synthesis, this type of reaction today belongs to the standard repertoire for the introduction of chiral alcohols by aldol-type reactions in natural product synthesis (*50–54*).

Although organocatalytic direct aldol reactions are very often carried out early in the synthesis of natural products, as shown in the case of epothilone B (**29**) (*49*) or in the synthesis of apratoxin A (*50*), there are also some excellent examples of late-stage enamine-catalyzed aldol reactions present in the literature (*51–54*). One good example refers to convolutamydine A (**32**), a naturally occurring potent inhibitor of HL-60 human promyelocytic leukemia cells (*55, 56*). Convolutamydine A (**32**) has been synthesized independently by several research groups (*51–54*) using a direct organocatalytic late-stage aldol reaction between acetone (**13**) and the dibromo-isatin **33**. These reports are remarkable for two reasons: (a) direct aldol reactions between two ketones are normally more difficult to execute than those with aldehyde acceptors due to the lower electrophilicity of ketones, and (b) all these reports used different amine-catalysts to achieve the same targeted transformation (Scheme 9).

Xiao et al. (*52*) used the bifunctional chiral bisamide **34** to catalyze the reaction between **13** and **33** to give (*S*)-**32** in a moderate enantiomeric excess of 60%. The enantiomeric excess (*ee*) could be enhanced significantly by a single crystallization (87%), albeit with a considerable decrease in yield. The enantioselectivity can be explained by the bifunctionality of catalyst **34** resulting in an enamine formation of the proline-nitrogen and acetone, accompanied with hydrogen bonds between the isatin carbonyl group and the two amide protons of the catalyst, leading to the correct orientation between electrophile and nucleophile.

Synthesis of the (*R*)-enantiomer of **32** was first accomplished by *Tomasini et al.* (*51*) using the proline amide catalyst **35**, resulting in an *ee* of 68% and excellent yield. In this case, **35** was superior when compared to the parent compound **12**, which displayed a poor enantioselectivity of less than 55% *ee* only (*51*).

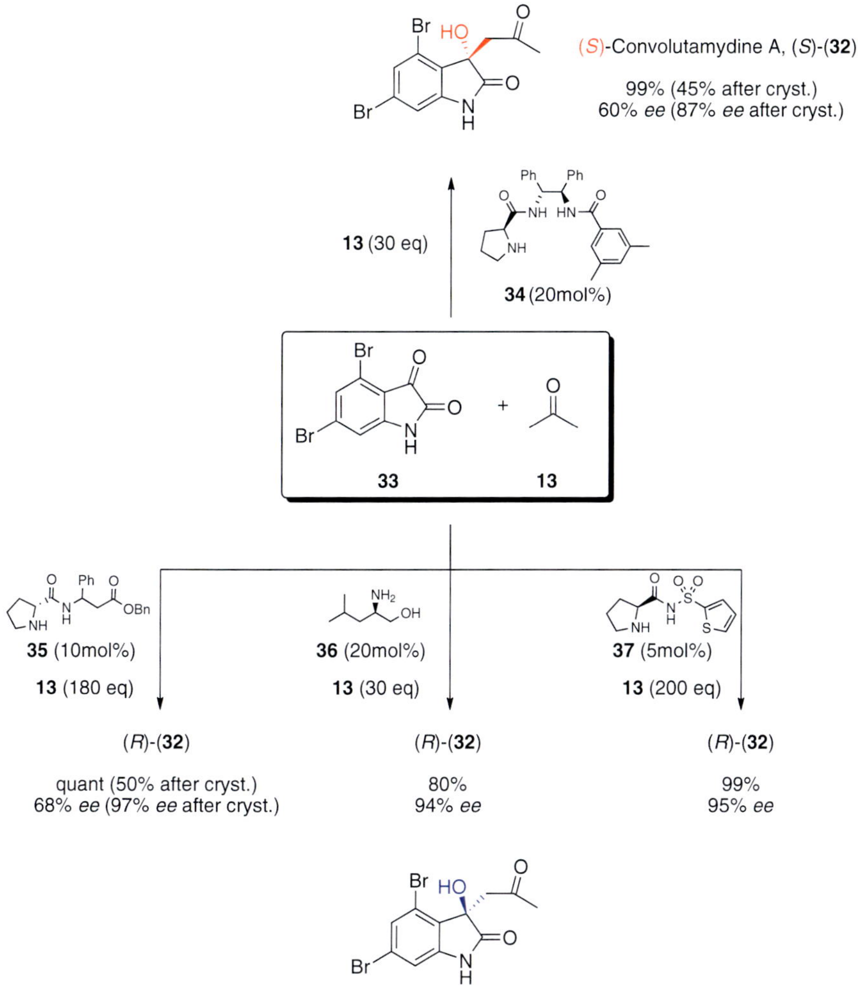

Scheme 9 Syntheses of (*R*)- and (*S*)-convolutamydine A (**32**)

Two high-yielding and highly enantioselective approaches were reported by *Malkov et al.* (*53*) and *Nakamura et al.* (*54*). Using D-leucinol (**36**) as the catalyst, *Malkov* and co-workers were able to obtain (*R*)-**32** in high yield and excellent enantioselectivity. Again, the high face selectivity can be rationalized by the presence of the hydroxy group of **36**, which is thought to coordinate the isatin keto group (*53*). Using only 5 mol% of the *N*-heteroarylsulfonylprolinamide catalyst **37**, *Nakamura et al.* were able to isolate (*R*)-**32** quantitatively in almost enantiopure form (*54*) (Scheme 9).

The high versatility of proline-catalyzed aldol reactions with ketone donors for the selective introduction of adjacent stereogenic centers was also applied

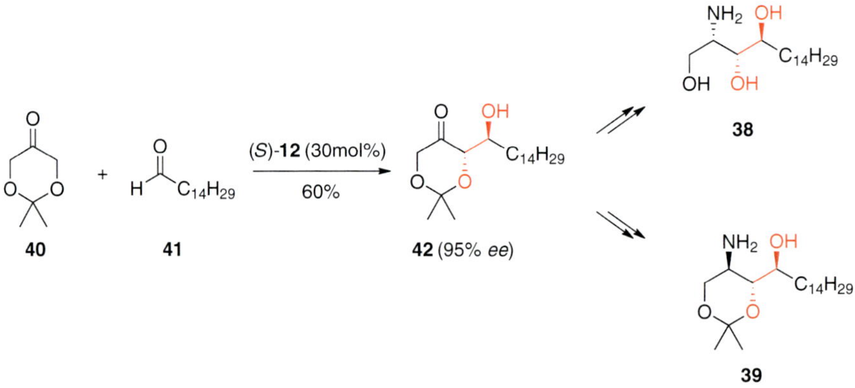

Scheme 10 Organocatalytic access to a key fragment in the synthesis of D-*arabino*-phytosphingosine (**38**) and protected L-*ribo*-phytosphingosine (**39**)

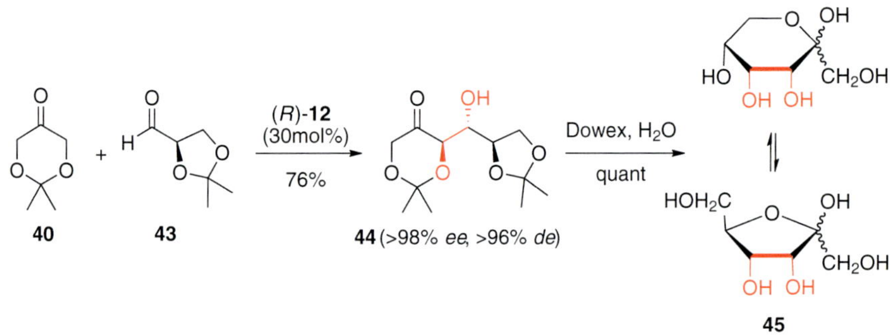

Scheme 11 *Enders*' carbohydrate synthesis (*57*)

successfully for the syntheses of carbohydrates and phytosphingosines, as demonstrated by *Enders et al.* (*57*, *58*). The short and flexible syntheses of D-*arabino*-phytosphingosine (**38**) and protected L-*ribo*-phytosphingosine (**39**) represent impressive examples for the successful application of this methodology (*58*). Herein, the characteristic amino-triol units of the sphingoids were introduced using the dioxanone **40** (*59*) and carrying out a proline-catalyzed aldol reaction giving the key fragment **42** in excellent stereoselectivity and good yield. Further functional group manipulations gave access to **38** and **39** in an easy and highly efficient manner (Scheme 10).

Using the dioxanone **40** as a synthetic dihydroxyacetone phosphate analogue, *Enders* and *Grondal* were able to synthesize several selectively protected carbohydrates in a direct and highly stereoselective fashion (*57*). As an example, the reaction between **40** and the aldehyde **43** catalyzed by (*R*)-**12** gave the acetonide-protected D-psicose **44** in 76% with excellent dia- and enantioselectivity. Deprotection of **44** gave D-psicose (**45**) quantitatively (Scheme 11).

Enders also applied a proline-catalyzed strategy to develop a direct biomimetically inspired route towards precursors of ulosonic and sialic acids (*60*) as demonstrated in

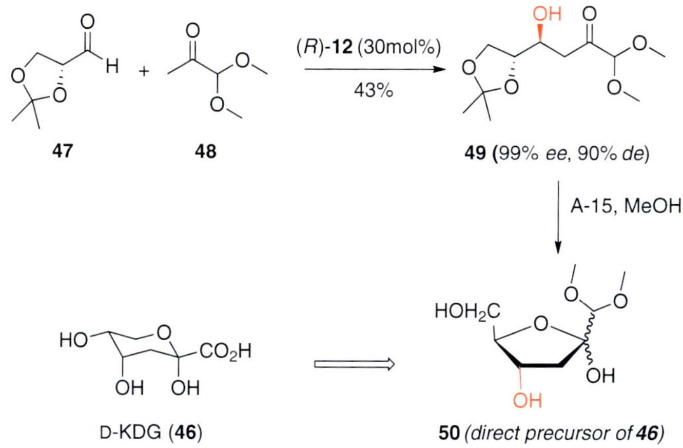

Scheme 12 Organocatalytic entry to a direct precursor of D-KDG (**46**)

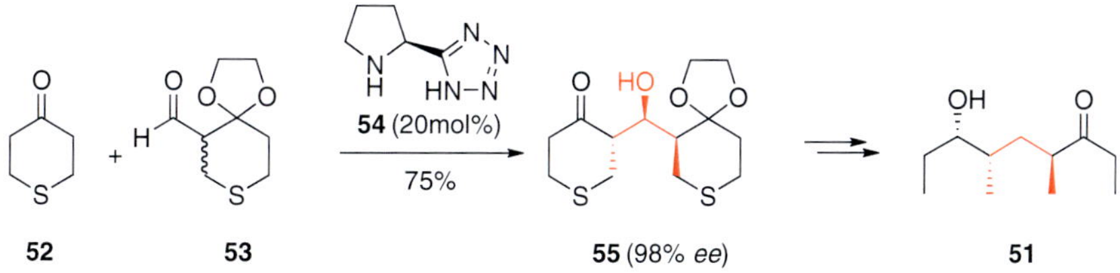

Scheme 13 Assembly of the tetrapropionate unit **55**, a key fragment in the synthesis of serricornin (**51**) (*62*)

the synthesis of a precursor of 2-keto-3-deoxy-D-glucosonic acid (D-KDG, **46**), a compound that takes part in the *Entner-Doudoroff* pathway in its phosphorylated form (*61*) (Scheme 12).

Another impressive short-step synthesis using this type of methodology was reported by *Ward et al.* (*62*). In their synthesis of serricornin (**51**), a sex pheromone produced by the female cigarette beetle (*Lasioderma serricorne*), the key step was an enantioselective aldol reaction between racemic aldehyde **53** and ketone **52** catalyzed by the tetrazole-catalyst **54** (*63–66*). This furnished the targeted tetrapropionate skeleton, which could be further transformed to the natural product **51** in six steps (Scheme 13). Interestingly, a concomitant dynamic kinetic resolution (DKR) of **53** was observed also under these conditions. It is worth noting that the key transformation can be carried out also in a highly selective manner but in a slightly lower yield using (*S*)-**12**. However, in this case a larger excess of ketone **52** was necessary, which complicated work up and purification on a larger scale (*62*).

Total synthesis of the potential anticancer drug salinosporamide A (**56**) represents an example where a proline-catalyzed aldol reaction between an achiral ketone donor and an α-chiral aldehyde was carried out with high selectivity.

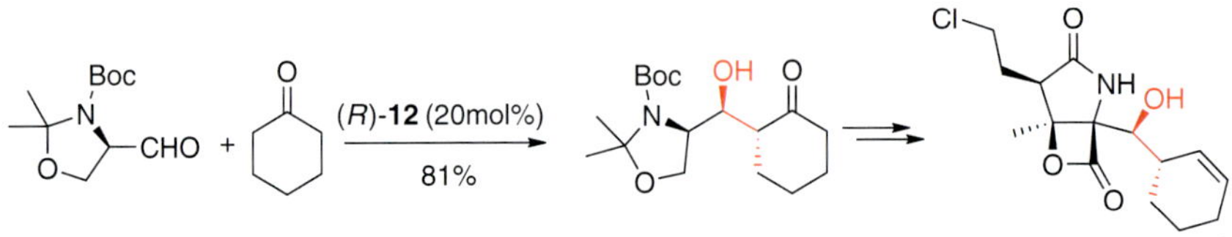

Scheme 14 Proline-catalyzed aldol reaction in the synthesis of salinosporamide A (**56**)

Reaction between cyclohexanone (**57**) and the α-chiral aldehyde **58** furnished the aldol product **59** in good yield and excellent diastereoselectivity (*67*, *68*) (Scheme 14).

2.1.2 Aldehyde Donors in Intermolecular Aldol Reactions

The first direct enantio- and diastereoselective organocatalytic cross-aldol reaction between aldehydes was reported by *MacMillan* in 2002 (*69*). Soon afterwards, *Pihko* and co-workers used this strategy for the reaction between propionaldehyde (**60**) and isobutyraldehyde (**14**). Carrying out this reaction under carefully controlled conditions with 10 mol% (*S*)-**12**, the β-hydroxy-aldehyde **61** could be obtained in good yield (61%) and excellent enantio- and diastereoselectivity (>99% *ee*, *dr* > 40:1) (*70*). Intermediate **61** could then be transformed readily into prelactone B (**62**), a natural product isolated from the bafilomycin-producing *Streptomyces griseus* (Scheme 15).

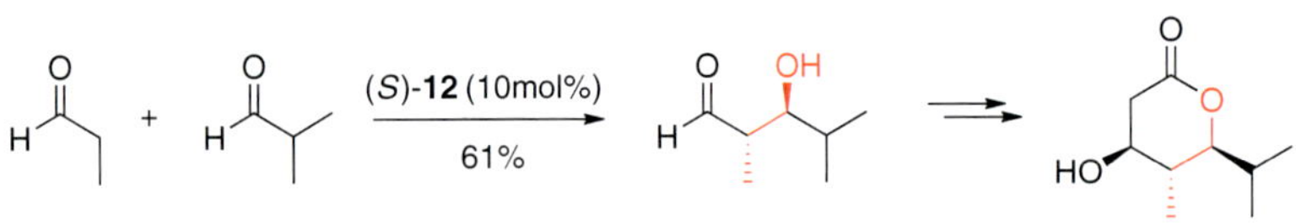

Scheme 15 Organocatalytic key transformation in the synthesis of prelactone B (**62**)

In 2004, the group of *MacMillan* published a report, which can now be regarded as one of the milestones in (organo-) catalysis. By carrying out an organocatalytic aldol reaction first, followed by a metal-catalyzed one, a two-step carbohydrate synthesis starting from simple starting materials could be achieved (*71*, *72*). The key organocatalytic step was a (*S*)-**12** catalyzed enantio- and diastereoselective dimerization of α-oxyaldehyde **63** (*72*). The α,γ-oxy-protected product **64** proved to be inert to further enolization or enamine addition. However, a subsequent *Lewis* acid-mediated aldol reaction of **64** with **65** then gave access to O-protected hexose

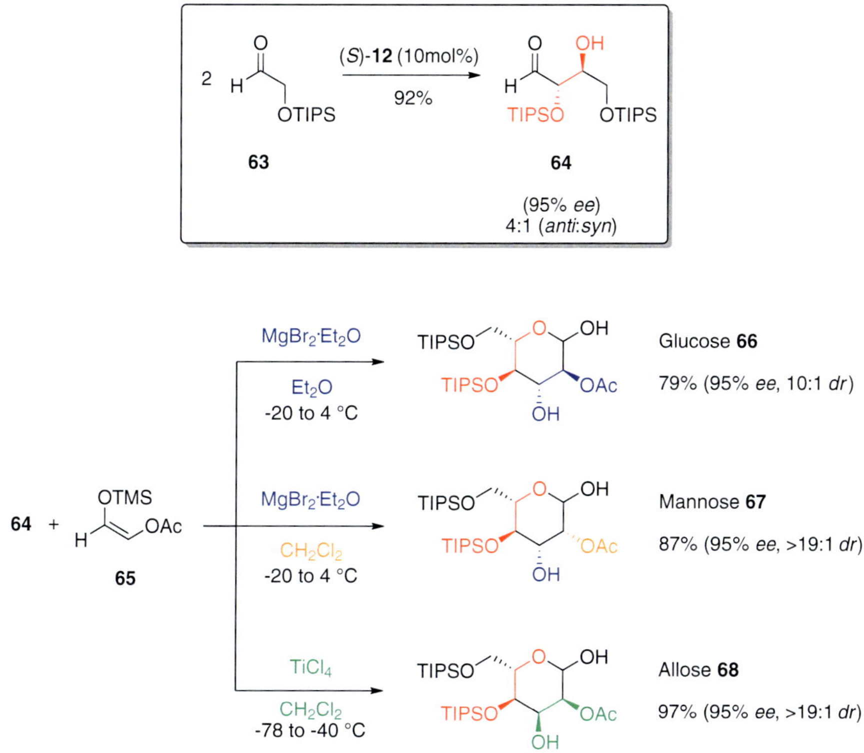

Scheme 16 *MacMillan's* two-step carbohydrate synthesis (*71*)

carbohydrates. It is worth noting that selective access to either protected glucose **66**, protected mannose **67**, or protected allose **68** can be accomplished by proper choice of the *Lewis* acid and the solvent (Scheme 16).

The spongistatin natural products (*e.g.* (+)-spongistatin 1 (**69**)) belong to some of the most potent antimitotic growth inhibitory substances discovered to date (*73–75*). However, their extremely low natural abundance (according to *Pettit* only 13.8 mg of (+)-spongistatin 1 (**69**) can be isolated from 400 kg of wet sponge (*73*)) fuels the demand for a synthesis approach to furnish sufficient amounts for further investigations. The group of *Amos B. Smith III* has for years been among the front-runners in the syntheses of complex natural products with a special focus on the development of scalable routes (*76–78*). Very recently, they reported an impressive gram-scale synthesis of **69** involving an early step organocatalytic aldol reaction to synthesize the F-ring (*78*). Synthesis of the F-ring started with an *anti*-selective aldol reaction between the TBDPS-protected electrophile **70** and propanal (**60**), in accordance with the procedure published by *MacMillan* (Scheme 16). Crude **71** was then submitted directly to an olefination reaction to give the key

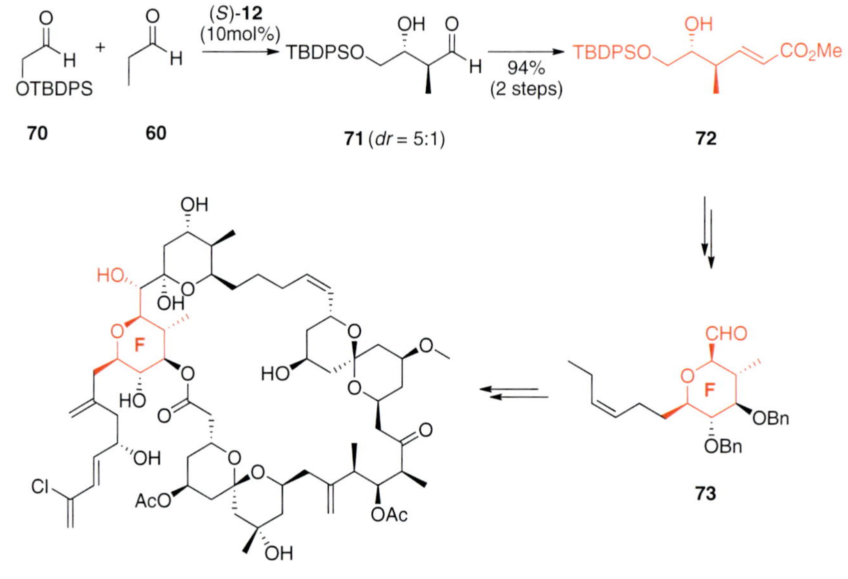

(+)-Spongistatin 1 (**69**)

Scheme 17 Organocatalysis in *A.B. Smith III's* gram-scale synthesis of (+)-spongistatin 1 (**69**)

intermediate **72**, which was used to synthesize the F-ring synthon **73**. Notably, the eight-step synthesis towards **73** proceeded with an impressive overall yield of 50%, which was significantly better than other approaches investigated in parallel by *Smith III et al*. Final assembly of the different synthons then gave access to over 1 g of (+)-spongistatin 1 (**69**) (Scheme 17) (*78*).

An aldehyde-aldehyde coupling was also the key transformation in the synthesis of the histone deacetylase inhibitor trichostatin A (**74**), developed by *Duan* and *Wang* (*79*). Reaction of *p*-nitrobenzaldehyde (**75**) and propionaldehyde (**60**) catalyzed by (*S*)-**12** gave the rather unstable enantiopure **76**, which was used directly for the subsequent steps towards trichostatin A (**74**). Notably, the chiral secondary alcohol group is oxidized later on in the sequence. Similar to some of the examples already depicted above, the organocatalytic introduction of the targeted stereogenic center is carried out very early in the multi-step sequence (Scheme 18) (*79*).

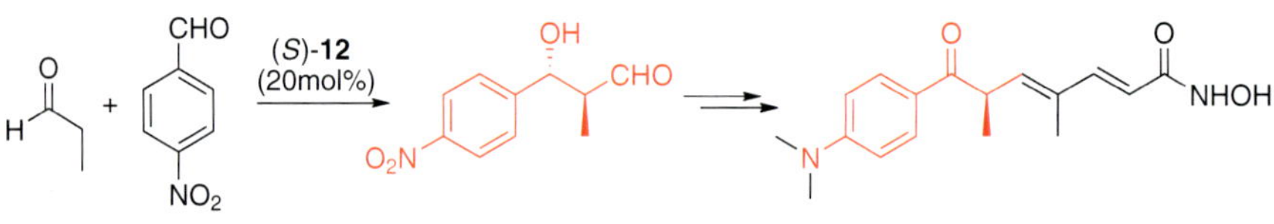

60 **75** **76** (>99% *ee*) Trichostatin A (**74**)

Scheme 18 Synthesis of trichostatin A (**74**)

Aldehyde donors were also employed successfully in the syntheses of convolutamydines E (**77**) and B (**78**) (*80–82*). The strategy was the same as depicted for the synthesis of (*R*)- and (*S*)-convolutamydine A (**32**) (Scheme 9), but using acetaldehyde (**79**) instead of acetone (**13**) as the nucleophile in the cross-aldol reaction with dibromo-isatin **33** (Scheme 19). *Nakamura et al.* utilized catalyst **37**, followed by a $NaBH_3CN$-mediated reduction to obtain (*R*)-convolutamydine E (**77**) in excellent yield and enantioselectivity. Chlorination of **77** then gave (*R*)-convolutamydine B (**78**) (Scheme 19) (*80, 81*).

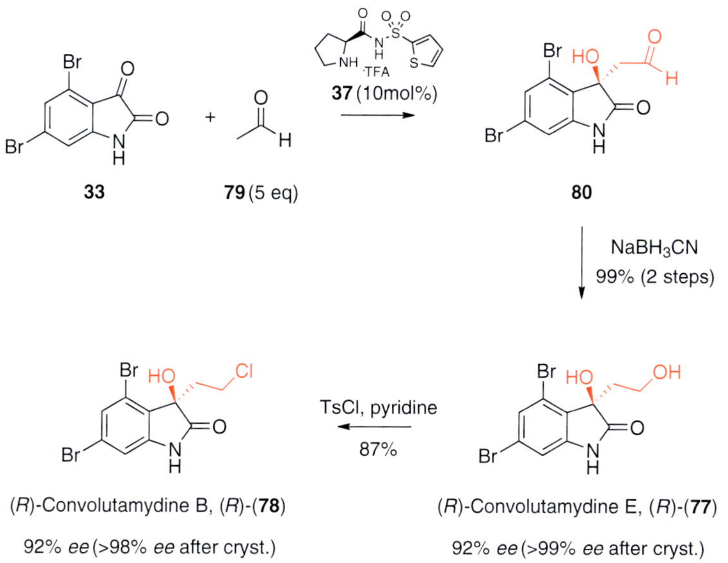

Scheme 19 *Nakamura's* syntheses of (*R*)-convolutamydine E (**77**) and B (**78**)

Synthesis of the unnatural (*S*)-convolutamydine E (**77**) was reported by *Hayashi et al.* using the diarylprolinol catalyst **81** (*82*). In contrast to other convolutamydine approaches, the N-protected isatins **82** and **83** were used, giving the aldol products **84** and **85** in good yield and *ee*. It is worth noting that in the case of **82** the (*S*)-configured **84** was obtained, whereas **83** gave the (*R*)-**85** under the same conditions. This difference in stereoselectivity was rationalized by *Hayashi et al.* as being due to the large C-5-substituent (Br) in the case of **82**, resulting in a different transition state than in the case of **83** with only a proton on C-5 (*82*). While **84** was easily converted to (*S*)-convolutamydine E (**77**), **85** could be used to synthesize either the alkaloid CPC-1 (**86**) (*83*), or to obtain **87**, a key fragment in the synthesis of madindolines A (**88**) and B (**89**), two selective inhibitors of interleukin-6, isolated from *Streptomyces nitrosporeus* K93-0711 (*84*) (Scheme 20).

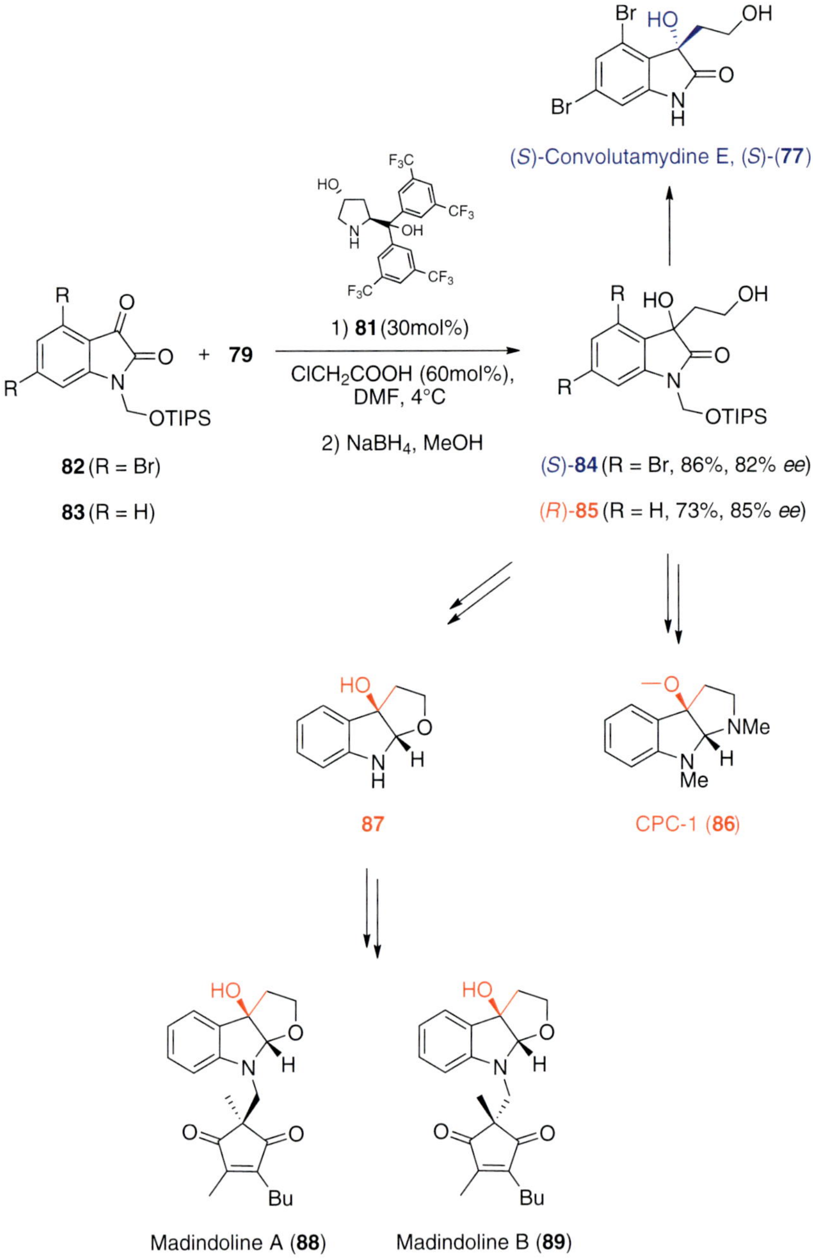

Scheme 20 Syntheses of (*S*)-convolutamydine E (**77**), CPC-1 (**86**), and a fragment (**87**) for the synthesis of madindolines A (**88**) and B (**89**)

2.1.3 Intramolecular Aldol Reactions

The *Hajos-Parrish-Eder-Sauer-Wiechert* synthesis (Scheme 5) was the first example of an intramolecular proline-catalyzed asymmetric aldol reaction. Systematically, this reaction can be described as a 6-*enolendo* cyclization. In 2003, *List et al.* described the first example of an intramolecular *enolexo* aldolization (*85*). This approach was then used by *Pearson* and *Mans* for the synthesis of (+)-cocaine **92**, starting from the *meso*-dialdehyde **90** on treatment with (*S*)-**12** (*86*). This desymmetrization process gave **91** as a mixture of epimers with good enantioselectivity. The tropane skeleton **91** could be further transformed into (+)-**92** by conventional means (Scheme 21).

Iwabuchi et al. developed an intramolecular desymmetrization approach for the conversion of the cyclohexanone **93** into the bicyclic **94** using the silylated hydroxyproline **95** as an enamine-catalyst (*87*). Using 25 mol% of **95**, the product **94** was obtained in good yield and excellent stereoselectivity and was employed

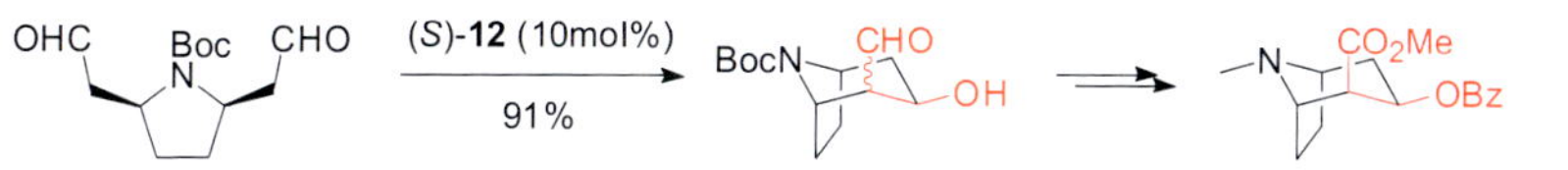

90 **91** (*dr* = 1:1) (+)-Cocaine (**92**) (86% *ee*)

Scheme 21 Organocatalytic desymmetrization approach towards (+)-cocaine (**92**)

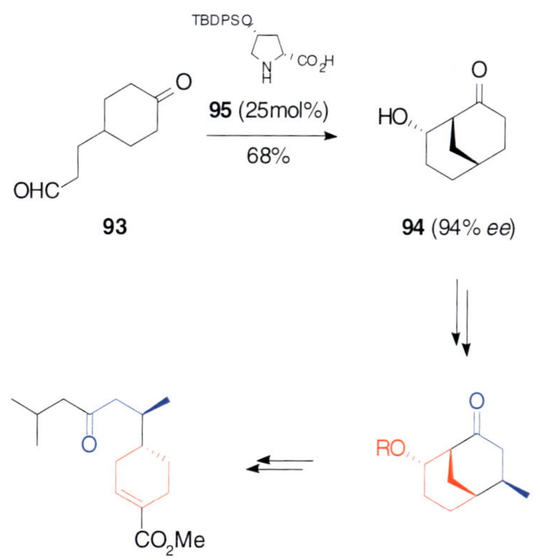

93 **94** (94% *ee*)

(+)-Juvabione (**96**) **97**

Scheme 22 Enamine-catalyzed synthesis of **94**, a key fragment in the synthesis of (+)-juvabione (**96**)

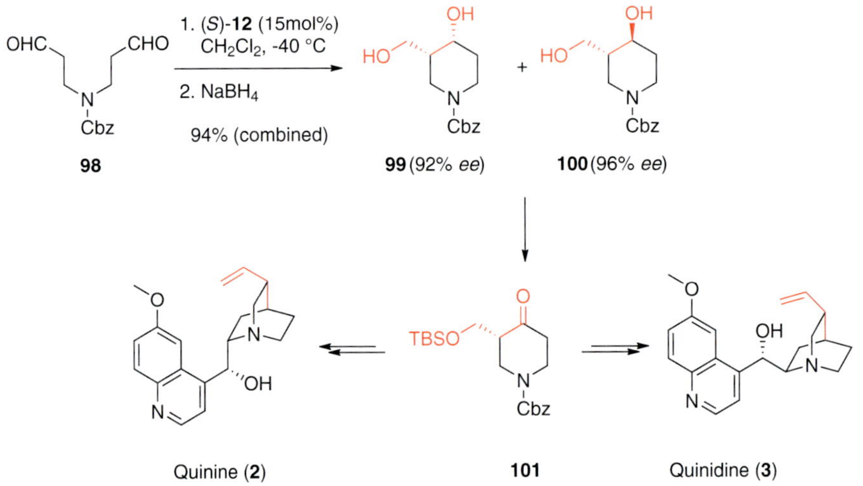

Scheme 23 Enamine-catalyzed introduction of one stereogenic center of quinine (**2**) and quinidine (**3**)

later on as a synthon for the synthesis of natural and non-natural targets (*88*, *89*). As an example of its usefulness as a starting material for complex natural compounds, *Iwabuchi et al.* developed an elegant synthesis of (+)-juvabione (**96**) (*89*), a natural sesquiterpene exhibiting insect juvenile hormone activity (*90*) (Scheme 22).

Very recently, an intramolecular cycloaldolization was employed successfully early in the synthesis of quinine (**2**) and quinidine (**3**) (*91*). *Hatakeyama et al.* used a (*S*)-**12** catalyzed aldol reaction followed by *in situ* reduction of the aldol product with $NaBH_4$ to obtain the diastereomers **99** and **100** in good yield and enantioselectivity (dr = 1:2). Followed by protection of the primary alcohol and oxidation of the secondary one, intermediate **101** with the desired configuration could easily be obtained. The intermediate **101** was then transformed into either **2** or the pseudoenantiomeric **3** by known methods (Scheme 23) (*91*).

2.2 *Mannich* Reactions

The *Mannich* reaction represents a useful extension of aldol-type approaches for the stereoselective formation of C–C bonds with concomitant introduction of O- and N-functionality. In this reaction two carbonyl compounds and an amine react to form β-amino-carbonyl compounds. Besides indirect variants using preformed enolates the use of unmodified nucleophiles (direct variant) has attracted considerable interest (*92*). Therefore, it is not surprising that the first example of a direct organocatalytic *Mannich* reaction was published only shortly (*93*) after the first proline-catalyzed aldol reaction (*28*). *p*-Anisidine (**102**) is commonly used for

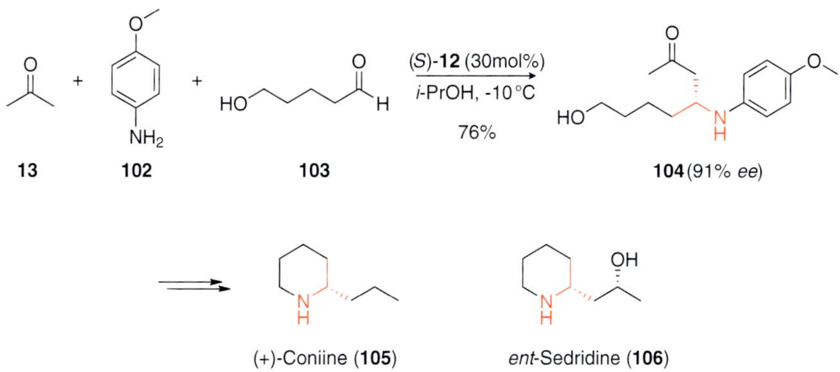

Scheme 24 Enamine-catalyzed three-component *Mannich* reaction in the syntheses of (+)-coniine (**105**) and *ent*-sedridine (**106**)

in situ imine formation and as an N-protecting group. However, its removal under strongly oxidizing conditions sometimes leads to incompatibilities. Thus, other protecting groups like Boc-groups also were used successfully in *Mannich*-type approaches. *Itoh et al*. developed a proline-catalyzed approach towards the intermediate **104** in the synthesis of (+)-coniine (**105**) and *ent*-sedridine (**106**) by carrying out a three-component *Mannich* reaction between **13**, **102**, and the hydroxy-aldehyde **103** (Scheme 24) (*94*, *95*).

Hayashi and co-workers used a similar strategy (Scheme 25) for the formal total synthesis of nikkomycin B (**107**) (*96*), a nucleoside peptide antibiotic isolated from the culture broth of *Streptomyces tendae*. In the key step, propionaldehyde (**60**), furfural (**108**), and the TBS-protected aniline **109** were reacted in the presence of

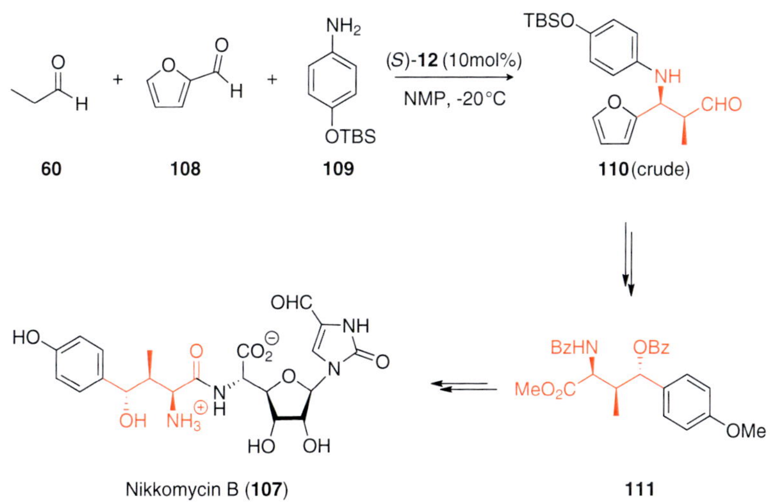

Scheme 25 Formal total synthesis of nikkomycin B (**107**)

10 mol% (*S*)-**12** to give the unstable chiral β-amino aldehyde **110** in high enantiopurity (determined later on in the sequence to be > 92%). Crude **110** was directly converted further representing a formal total synthesis of **107** (*96*).

Besides three-component direct *Mannich* approaches, also the strategy of using preformed imines was applied successfully. For example, a preformed Boc-protected imine was used successfully by the *Enders* group in their synthesis of (+)-polyoxamic acid (**112**) (*97*). Polyoxamic acid is one of the saponification fragments of polyoxin J (**113**), an unusual peptidyl nucleosidic antibiotic isolated from the culture broths of *Streptomyces cacoi* var. *asoensis* (*98*). The preformed Boc-imine of furfural (**114**) was reacted with the dioxanone (**40**) in the presence of (*S*)-**12**. The amino ketone **115** could be obtained in good yield and enantioselectivity and was easily transformed into (+)-polyoxamic acid (**112**) (Scheme 26) (*97*).

A three-component *Mannich* reaction between *p*-methoxybenzaldehyde (**116**), hydroxyacetone (**117**), and **102** was used successfully to build up the two consecutive stereogenic centers of **118**, an intermediate in the synthesis of (+)-*epi*-cytoxazone (**119**), as shown by *Sudalai et al.* (*99*) (Scheme 27). Cytotoxazone was isolated from *Streptomyces* sp. and is a potent inhibitor of the signaling

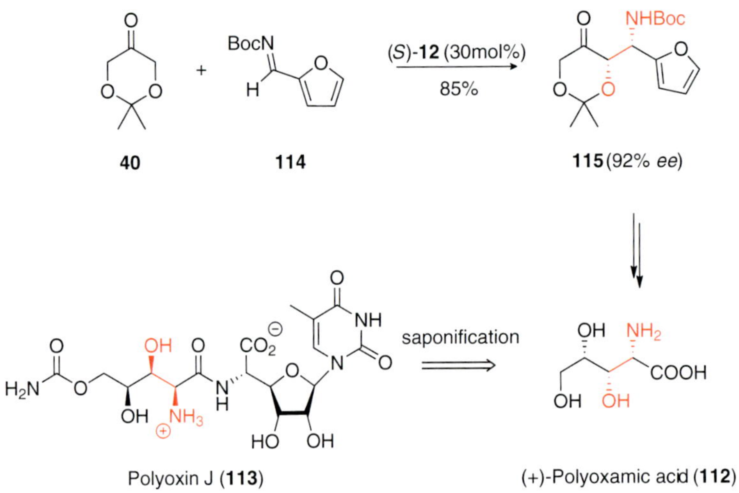

Scheme 26 *Mannich* reaction in the synthesis of (+)-polyoxamic acid (**112**)

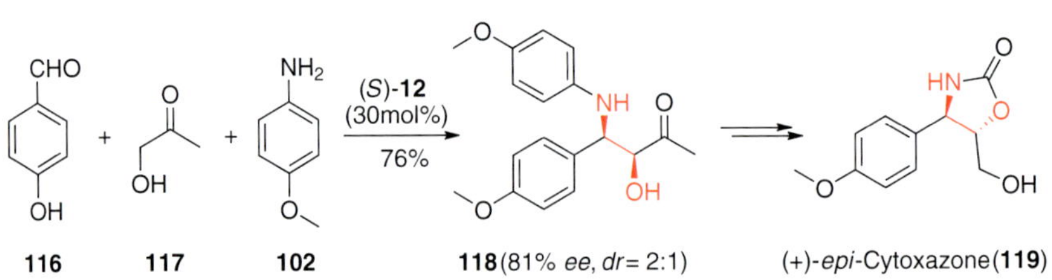

Scheme 27 *Mannich* reaction in the synthesis of (+)-*epi*-cytoxazone (**119**)

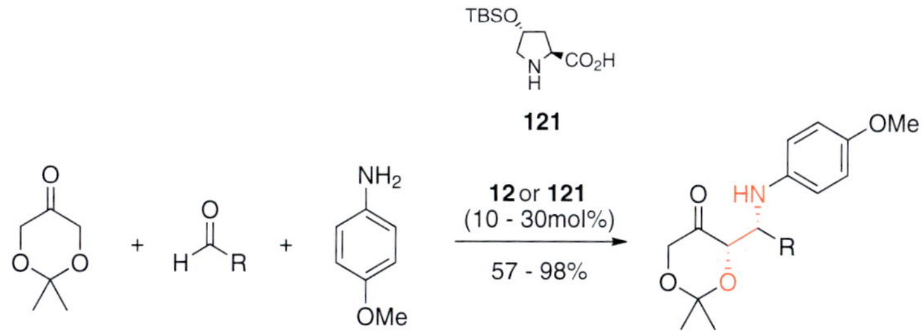

40 120 102 122 (81 - 98% *ee*, 78 - 99% *de*)

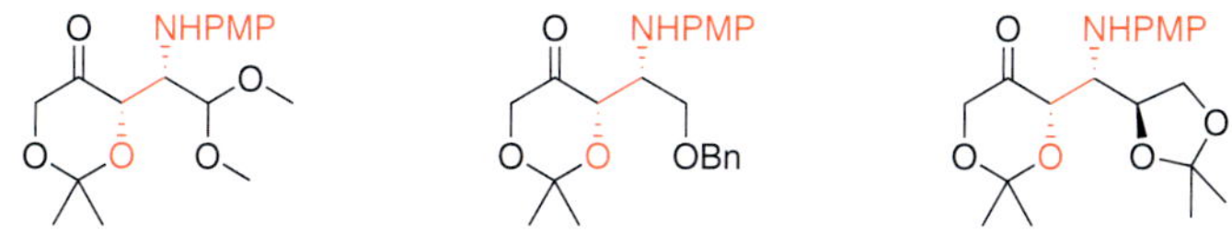

122a protected 2-amino-2-deoxy-D-*threo*-pentos-5-ulose

122b protected 4-amino-4-deoxy-D-xylulose

122c protected 4-amino-4-deoxy-D-fructose

Scheme 28 Enamine-catalyzed three-component direct *Mannich* reaction for the syntheses of protected amino sugars

pathways of Th2 cells and therefore a potential chemotherapeutic agent in the field of immunotherapy (*100*).

By analogy to their elegant approach for the syntheses of carbohydrates (Scheme 11) (*57*) and phytosphingosines (Scheme 10) (*58*), *Enders* and co-workers used their dioxanone **40** (*59*) to synthesize a variety of protected amino sugars by a direct organocatalytic *Mannich* reaction (*101*). This approach is especially interesting as the syntheses of aminosugars are normally rather challenging including several protecting group manipulations, carbon-carbon bond formations and oxidation-reduction steps (*102–104*). In contrast, reacting **40** with different aldehydes **120** and the aniline **102** in the presence of either (*R*)- or (*S*)-proline (**12**) or the protected hydroxyproline (**121**) gave access to several protected amino pentoses and hexoses with high diversity in just one step (Scheme 28) (*101*).

2.3 α-Heterofunctionalizations

Introducing heteroatoms in the α-position of carbonyl groups in a direct approach is one of the types of transformations that have benefitted the most from recent progress in the field of enamine catalysis (*105–109*). Besides the development of suitable methods for the stereoselective introduction of O- and N-heteroatoms, α-halogenation approaches have attracted considerable interest over the last few years (*36*). However, whereas stereoselective α-amination and α-oxygenation strategies have been used in several natural product and natural product analogue syntheses, enamine-catalyzed α-halogenations, especially α-fluorinations, were used mainly in the syntheses of non-natural products. Therefore, the main focus in this chapter will be on the introduction of N- and O-heteroatoms.

2.3.1 α-Hydroxylation

One of the requirements for such reactions is the availability of suitable electrophilic reagents for the introduction of heteroatoms. Nitrosobenzene (**123**) is normally the reagent of choice for the α-oxygenation of carbonyl groups, giving α-anilinooxy carbonyl compounds as the primary products. The use of **123** benefits from its high reactivity and the fact that the N-O bond easily can be cleaved (*e.g.* Cu (II)-salts) to give the corresponding chiral α-hydroxy carbonyl compounds (*110*).

An illustrative example for the strength of this methodology was reported by *Kim et al.* who used an enamine-catalyzed α-oxidation in the synthesis of the bark beetle pheromones (+)-*exo*- and (−)-*endo*-brevicomin (**124**) (*111*). The asymmetric proline-catalyzed α-hydroxylation of butyraldehyde (**125**) with **123** gave the crude anilinooxy compound **126** in excellent enantioselectivity (>98%). Compound **126** was used directly further to give (+)-*exo*- and (−)-*endo*-brevicomin (**124**) in just three more steps (Scheme 29) (*111*).

Hayashi et al. used (*R*)- or (*S*)-proline as catalysts to hydroxylate the cyclohexanone **127** selectively to get either the intermediate **128** (upon hydrogenolysis of the N—O bond of the primary reaction product), which was converted further to the natural anti-angiogenesis agent RK-805 (**129**) and similar natural products such as fumagillol and ovalicin (*112*), or the intermediate **130**. The latter could be transformed into (+)-panepophenanthrin (**131**) (*113*) (Scheme 30), a potent

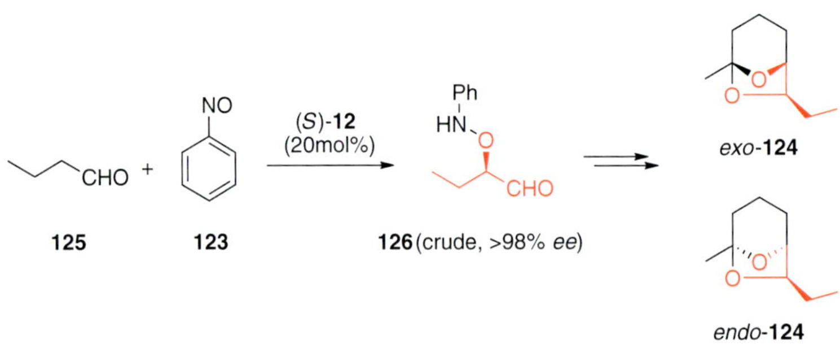

Scheme 29 Proline (**12**)-catalyzed α-oxygenation in the synthesis of (+)-*exo*- and (−)-*endo*-brevicomin (**124**)

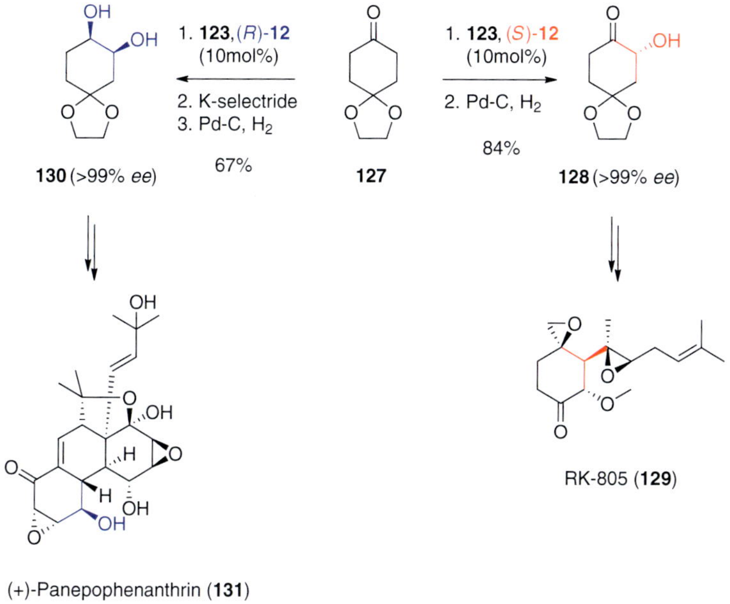

Scheme 30 Proline (**12**)-catalyzed α-hydroxylation for the synthesis of RK-805 (**129**) and (+)-panepophenanthrin (**131**)

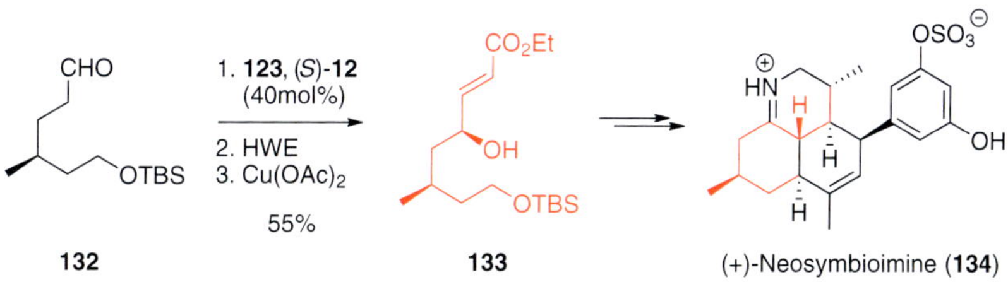

Scheme 31 Total synthesis of neosymbioimine (**134**)

inhibitor of the ubiquitin-activating enzyme (E1) that plays an important role in activating the ubiquitin-proteasome pathway (*114*).

Maier and *Varseev* recently applied *MacMillan's* elegant one-pot three-step protocol (*115*) (1. α-hydroxylation, 2. *Horner-Wadsworth-Emmons* (*HWE*) olefination, 3. N—O cleavage) for the conversion of the (*S*)-citronellol-derived aldehyde **132** to γ-hydroxy-α,β-unsaturated ester **133** early in their 18-step first total synthesis of neosymbioimine (**134**) (*116*), a minor amphoteric metabolite of the symbiotic marine dinoflagellate *Symbiodinium* sp. (*117*) (Scheme 31).

The synthesis of disparlure (**135**) represents yet another fine example for the high potential of enamine-catalyzed α-functionalizations of aldehydes in the synthesis of natural products. The enantiomer (+)-**135** is a sex pheromone of the female gypsy

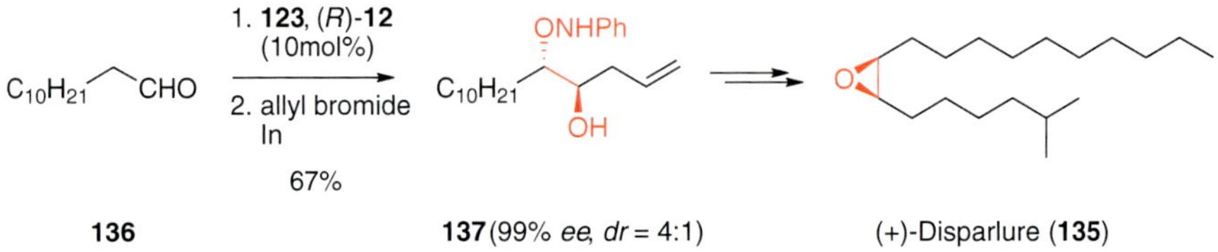

Scheme 32 Total synthesis of (+)-disparlure (**135**)

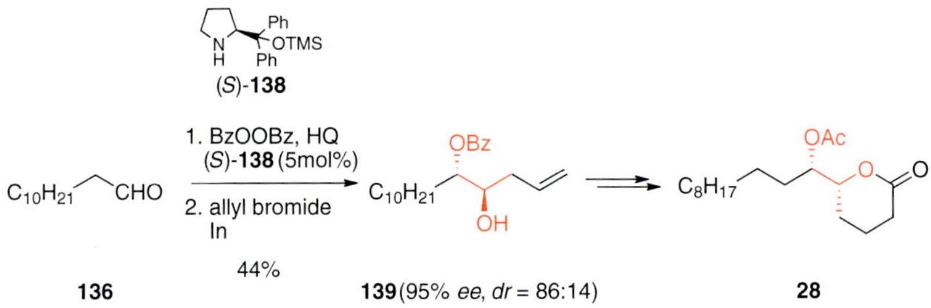

Scheme 33 Enamine-catalyzed α-oxygenation in the synthesis of (−)-(5*R*,6*S*)-6-acetoxy-5-hexadecanolide (**28**)

moth, *Porthetria dispar* (*118*), whereas the (−)-enantiomer has been shown to antagonize this effect (*119*). *Kim* reported a proline-catalyzed approach towards both enantiomers (*120*). Thus, the aldehyde **136** was aminooxylated with **123** in the presence of catalytic amounts of (*R*)-**12** followed by direct allylation of the primary reaction product to give the homoallylic alcohol **137** in good yield, good diastereoselectivity and excellent enantioselectivity. Standard manipulations then gave (+)-disparlure (**135**) in a total of six steps with 30% overall yield (Scheme 32). Using (*S*)-**12** as a catalyst for the α-oxygenation gave access to (−)-**135** in the same way.

As depicted in Scheme 7, the synthesis of the mosquito oviposition pheromone (-)-6-acetoxy-5-hexadecanolide (**28**) *via* an intermolecular aldol reaction represents a powerful demonstration of the high potential of asymmetric enamine catalysis (*45*, *46*). It is noteworthy that a methodologically different successful organocatalytic approach towards **28**, based on an asymmetric α-oxygenation, was reported recently (*121*). Reaction of aldehyde **136** with dibenzoyl peroxide (BzOOBz) and hydroquinone (HQ) (*122*) in the presence of the TMS-protected prolinol catalyst (*S*)-**138** followed by a direct allyation gave the benzoyl-protected **139** in moderate yield and good selectivity. Intermediate **139** could then be further transformed to give (−)-(5*R*,6*S*)-6-acetoxy-5-hexadecanolide (**28**) (Scheme 33).

2.3.2 α-Amination

The direct stereoselective introduction of a nitrogen functionality in the α-position of a carbonyl group is of high interest as it leads to valuable compounds like

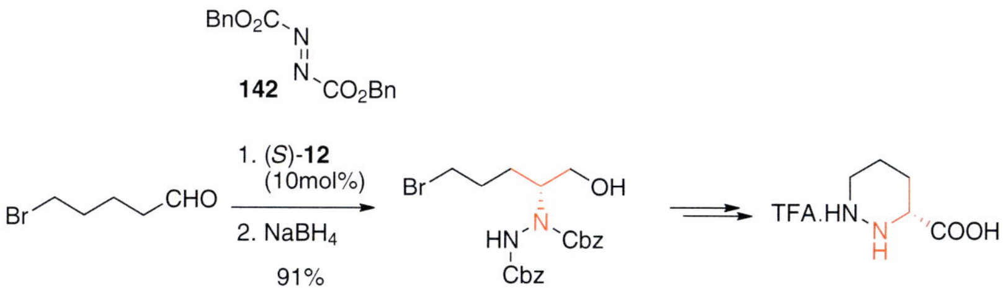

Scheme 34 Asymmetric α-nitrogen-functionalization in the synthesis of (*R*)-piperazic acid (**140**)

α-amino acids, α-amino aldehydes, and β-amino alcohols, and it also gives access to several intermediates in complex natural product syntheses (*123–127*). The first direct catalytic α-aminations of aldehydes were reported independently by *Jørgensen et al.* (*123*, *124*) and *List et al.* (*125*), both using azodicarboxylates as electrophilic N-agents.

The cyclic α-hydrazinocarboxylic acids (*R*)- and (*S*)-piperazic acid (**140**) are present in a variety of bioactive cyclodepsipeptides. *Hamada et al.* have developed a highly efficient synthesis of (*R*)-(**140**) by reacting 5-bromovaleraldehyde (**141**) with azodicarboxylate **142** in the presence of 10 mol% (*S*)-**12** (*128*). The reaction product was directly treated with $NaBH_4$ to give the alcohol **143** in excellent yield and enantioselectivity. The product **140** could then be obtained successfully after cyclization, oxidation, and Cbz-cleavage (Scheme 34).

Lycopodium alkaloids have always been attractive targets for synthesis-oriented organic chemists due to the combination of having unique polycyclic structures with promising biological activities (*129–135*). A few years ago, *Kobayashi et al.* isolated cermizine C (**144**), cermizine D (**145**), and senepodine G (**146**) together with other alkaloids from the club mosses *Lycopodium cernuum* and *L. chinense* (*136*). It is of interest to note that whereas **145** and **146** exhibited *in vitro* cytotoxicity against murine lymphoma L1210 cells (IC_{50} 7.5 and 7.8 μg/cm^3), **144** did not show cytotoxicity at 10 μg/cm^3. In contrast to the relatively recent discovery of **144–146**, the structurally related cernuane-type alkaloid cernuine (**147**) was isolated over 60 years ago by *Marion* and *Manske* (*137*). The *Takayama* group has for several years been at the forefront in the synthesis of cernuane- and quinolizidine-type alkaloids (*138–142*). Recently, they developed a divergent strategy for the syntheses of **144–147** starting from the (+)-citronellal-derived aldehyde **148** (*138*, *139*). Organocatalytic α-amination of **148** catalyzed by (*R*)-**138** gave a rather labile amination product that was directly converted further to the stable oxazolidinone **149**. Further modification then gave the key intermediate **150**, which could be used to obtain cermizine C (**144**), cermizine D (**145**), and senepodine G (**146**), as well as cernuine (**147**) (Scheme 35).

It should be noted that stereoselective enamine-catalyzed *a*-functionalizations have also been used very often in combined organocatalytic approaches, *e.g.*

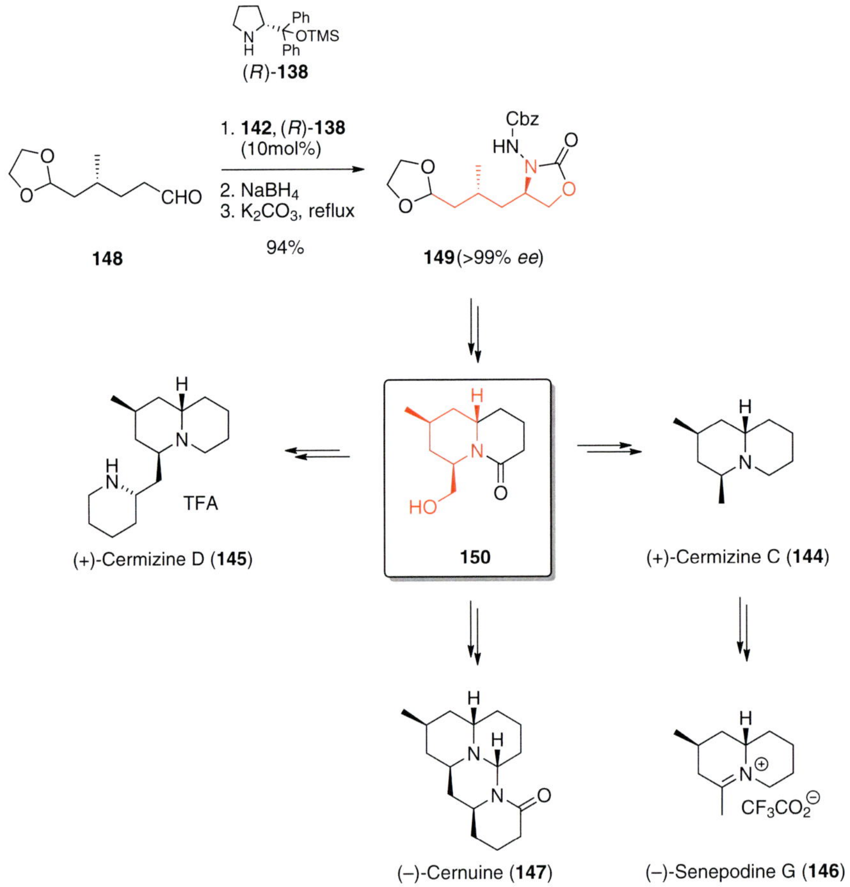

Scheme 35 Asymmetric α-amination in the synthesis of the key fragment **150** for the syntheses of the alkaloids cermizine C (**144**), cermizine D (**145**), and senepodine G (**146**), and cernuine (**147**)

accompanied with conjugate additions or other organocatalytic transformations. Some examples, therefore, will be covered in a later section (see Sect. 2.6).

2.4 Conjugate Additions

Conjugate additions of enamines to α,β-unsaturated carbonyl or nitro compounds belong to the most commonly applied and useful organocatalytic C–C bond-forming reactions (*135, 143–150*). As already mentioned above, enamine-catalyzed conjugate additions are often used in combination with α-heterofunctionalizations. These combined examples will be described in more detail in Sect. 2.6.

Besides the use of enamine-catalyzed *Michael*-type reactions in natural product syntheses, also its high potential for the syntheses of bioactive "designer drugs" has been demonstrated. As an example, *Hayashi et al.* developed an efficient, enantioselective total synthesis of the anti-influenza neuramidase inhibitor (-)-oseltamivir using an enamine-catalyzed *Michael* addition early in the sequence (*151*). This strategy was especially interesting, as it provided a novel, non-shikimic acid-based approach towards (−)-oseltamivir phosphate (Tamiflu®), one of the most prominent antiviral drugs that is currently on the market (*152*). As this impressive example is not part of a natural product total synthesis, no further details will be given in this review but the interested reader may be referred to the original literature (*151*).

Similar to aldol approaches, also organocatalytic conjugate addition reactions have been used mainly in the early steps of complex natural product syntheses to obtain useful chiral synthons in an easy and reliable fashion. Very recently, *Carter et al.* described the application of an intramolecular highly diastereoselective *Michael* addition of a chiral starting material catalyzed by an achiral amine for the total synthesis of lycopodine (*129*, *135*). Accordingly, there is not only a high demand for enamine catalysis in the syntheses of chiral molecules using enantiopure chiral amines, but also the use of achiral amines for the diastereoselective transformation of chiral starting materials is highly appreciated.

Interesting examples for conjugate additions mediated by chiral amines have been described by *Alexakis et al.* (Scheme 36), who used the nitroalkene **151** as a *Michael* acceptor in organocatalytic enamine-catalyzed conjugate addition reactions (*149*, *150*, *153*). *Michael* reaction of **151** with propionaldehyde **60** in the presence of the diamine catalyst **152** (15 mol%) gave **153** as a mixture of four diastereomers in good yield. Subsequent aldehyde protection and conversion of the

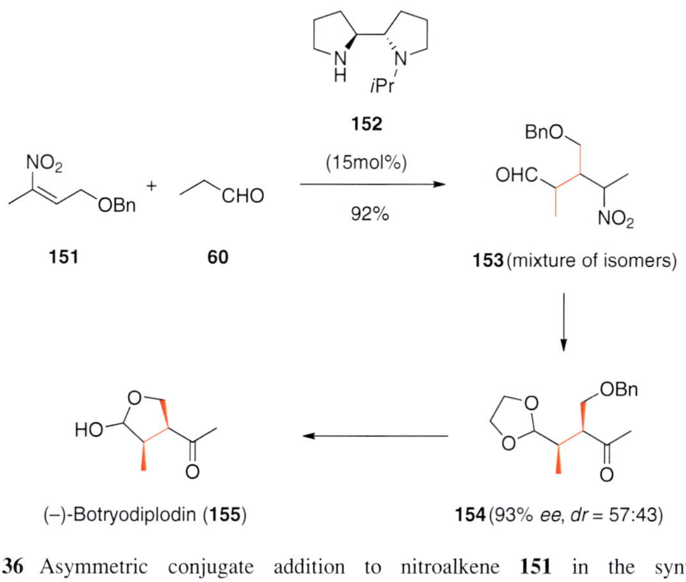

Scheme 36 Asymmetric conjugate addition to nitroalkene **151** in the synthesis of (−)-botryodiplodin (**155**)

nitro group into a keto group resulted in **154** as a mixture of two diastereomers in good enantioselectivity. The major isomer was utilized successfully to synthesize (−)-botryodiplodin (**155**) (*153*), an antibiotic isolated from the plant pathogen *Botryodiplodia theobromae*, which causes many tropical fruit diseases including mango twig blight and mango stem rot (*154*).

Recently, the group of *Nicolaou* (*155*) reported a very elegant 12-step enantioselective synthesis of biyouyanagin A (**156**), an anti-HIV-active compound isolated from *Hypericum chinense* var. *salicifolium,* which is used as a common folk medicine in Japan (*156*). As already shown in several examples so far, organocatalysis was found to be rather fruitful early on in this synthesis, giving access to a key synthon in a highly stereoselective manner. The sequence started with a *Michael* addition of (*R*)-citronellal (**157**) to methyl vinyl ketone (**158**), catalyzed by 5 mol% of (*S*)-**159** and 20 mol% of catechol **160**, which is considered to function as a co-catalyst *via* hydrogen bond donation to the enone (*157*). The reaction was carried out as an organocatalytic cascade reaction, proceeding *via* the primary addition product **161**, which was directly transformed further into the enone **162** by an intramolecular aldol condensation. Employing this strategy, **162** could be obtained in 68% yield and reasonable stereoselectivity (*de* = 86%). Further manipulations then accomplished the total synthesis of biyouyanagin A (**156**) in a straightforward and effective way (Scheme 37) (*155*).

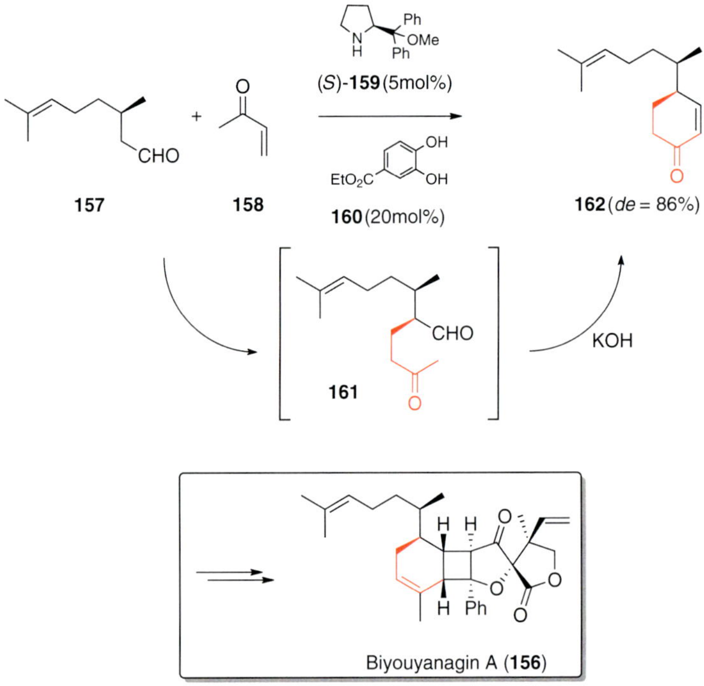

Scheme 37 Early stage enamine-catalyzed *Michael* addition in *Nicolaou's* total synthesis of biyouyanagin A (**156**)

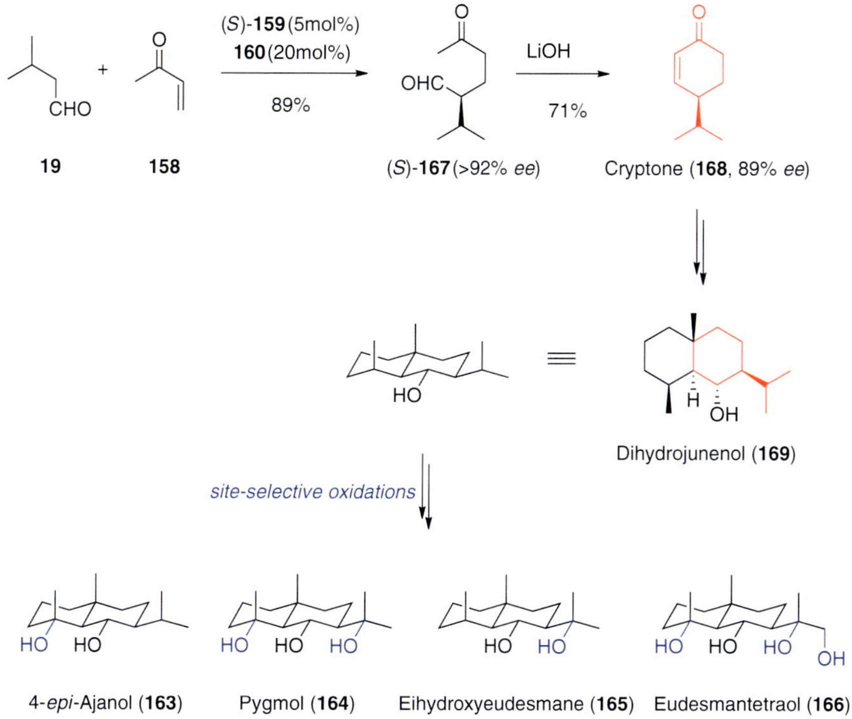

Scheme 38 Enamine-catalyzed *Michael* addition for the synthesis of the eudesmane terpene precursor **168**

A rather similar organocatalytic strategy was applied by *Chen* and *Baran* in the first steps of their syntheses of the eudesmane terpenes **163–166** (*158*). The eudesmane family of sesquiterpenoids contains over 1000 members with almost every conceivable oxidation pattern expressed (*159*). Despite their low molecular weight, the rigid skeletons of these natural products make them difficult targets for synthesis. *Baran* and *Chen* developed a scalable procedure towards these interesting compounds. By analogy to *Nicolaou's* approach (Scheme 37), the sequence commenced with a *Michael* addition of isovaleraldehyde (**19**) to enone **158**. The primary *Michael* product (*S*)-**167** was directly cyclized to afford the natural product cryptone (**168**), which was then transformed into dihydrojunenol (**169**) on a multigram scale in seven steps. The substituted *trans*-decalin **169** then served as the substrate of choice for a variety of very impressive site-selective C-H functionalization reactions, which gave access to the four eudesmane terpenes **163–166** in good yields and excellent selectivities (Scheme 38) (*158*).

Finally, the high versatility of this *Michael* strategy was also shown in the total synthesis of the antimalarial agent (+)-polyanthellin A (**170**), which was described recently by *Johnson et al.* (*160*). The first step was similar to that depicted in Scheme 38, only using (*R*)-**159** as a catalyst for the *Michael* reaction between

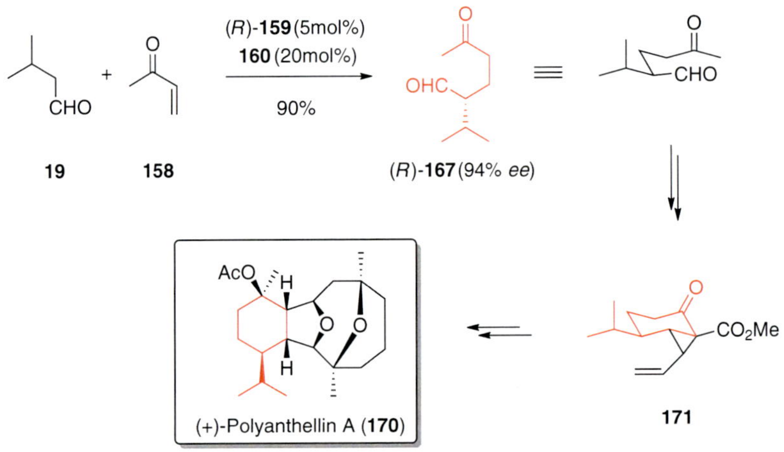

Scheme 39 Enamine-catalyzed *Michael* addition in the synthesis of (+)-polyanthellin A (**170**)

isovaleraldehyde (**19**) and enone **158**. In this case, the (*R*)-configured **167** could be obtained in excellent yield and enantioselectivity. In contrast to the previous examples, no intramolecular aldol condensation was carried out but **167** was used to synthesize the bicyclic compound **171**, which served as one of two key building blocks for the final approach towards (+)-polyanthellin A (**170**) (Scheme 39) (*160*).

Once again it should be noted that the main purpose of this volume is to illustrate the high potential of organocatalysis and therefore the focus on the organocatalytic transformations in all these multi-step sequences. Very often these reactions have been carried out early in a sequence, giving access to chiral precursors and key intermediates in an elegant and easy fashion. Therefore, some of the most highly innovative transformations, which were applied later in the final steps, are not given in detail, as this is beyond the scope of this account.

Recently, the dual-specifity phosphatase inhibitor (−)-bitungolide F (**172**) was synthesized successfully by the *Cossy* group (*161*). The nine-step synthesis started with an organocatalytic *Michael* addition of aldehyde **60** to methyl vinyl ketone (**158**). The enantio-enriched **173** was then used to synthesize (−)-bitungolide F (**172**) in a straightforward manner and good yield (nine steps, 11% overall yield) (Scheme 40) (*161*).

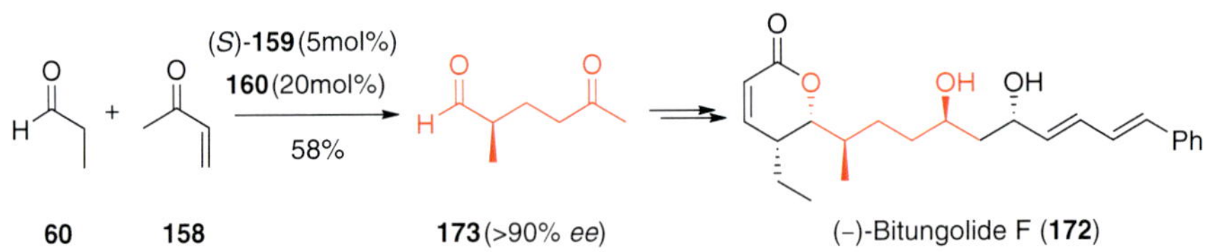

Scheme 40 Enamine-catalyzed *Michael* addition in the synthesis of (−)-bitungolide F (**172**)

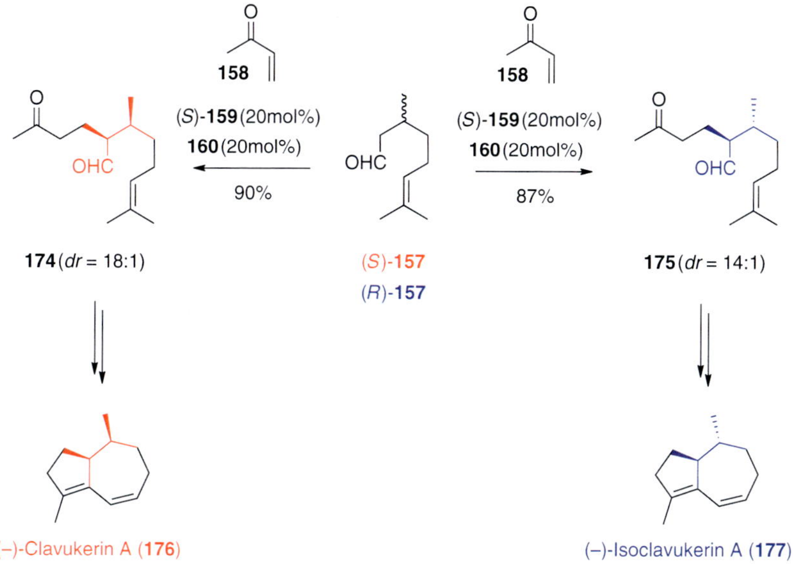

Scheme 41 *Michael* addition in the syntheses of (−)-clavukerin A (**176**) and (−)-isoclavukerin A (**177**)

The group of *Metz* employed the proline-derived catalyst **159** in combination with co-catalyst **160** to catalyze additions of (*R*)- and (*S*)-citronellal (**157**) to **158** for the selective syntheses of the diastereomeric keto aldehydes **174** and **175**. These intermediates could then be used to synthesize the marine sesquiterpenoids (−)-clavukerin A (**176**) (starting from (*S*)-**157**) and (−)-isoclavukerin A (**177**) (derived from (*R*)-**157**) (Scheme 41) (*162*).

Anominine (**178**) is an indole terpene (Scheme 42) isolated from the sclerotia of *Aspergillus nomius* by *Gloer et al.* (*163*). The natural (+)-**178** exhibits potent activity against the widespread crop pest *Heliothis zea* in controlled feeding experiments. Very recently, the (−)-enantiomer was synthesized successfully by the group of *Bonjoch* for the first time (*164*). This impressive route commenced

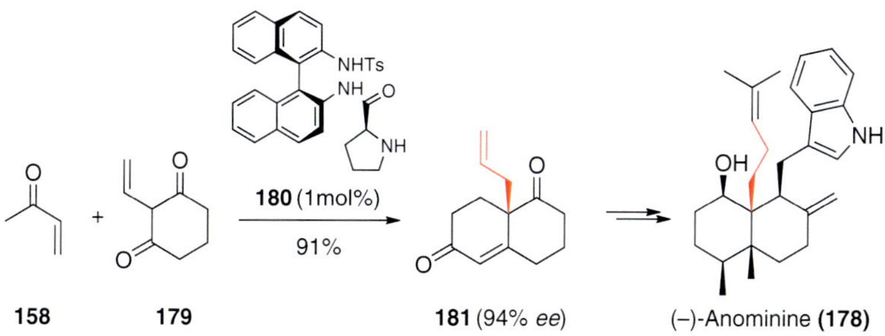

Scheme 42 Enamine-catalyzed *Robinson* annulation in the synthesis of (−)-anominine (**178**)

with a stereoselective *Robinson* annulation between **158** and **179** catalyzed by only 1 mol% of the binaphthyl-derived catalyst **180** (*165*) to build up the *Wieland-Miescher* ketone **181**. Compound **181** was then used successfully to achieve the first total synthesis of (−)-anominine (**178**) in several steps (*164*).

2.5 Dienamine Catalysis

The use of an amine catalyst for the activation of an α,β-unsaturated carbonyl group as a nucleophile represents a very useful extension of the concept of enamine activation. By transmission of the nucleophilic properties of an enamine to an adjacent olefin (vinylogy), the dienamine can be considered to be a compound bearing two nucleophilic sites (the α- and the γ-position) and furthermore it can react as an electron-rich diene in [4 + 2] cycloadditions. This concept was introduced in 2006 by *Jørgensen et al.* who successfully applied it for the stereoselective γ-amination of α,β-unsaturated carbonyl compounds (*166*) (Scheme 43).

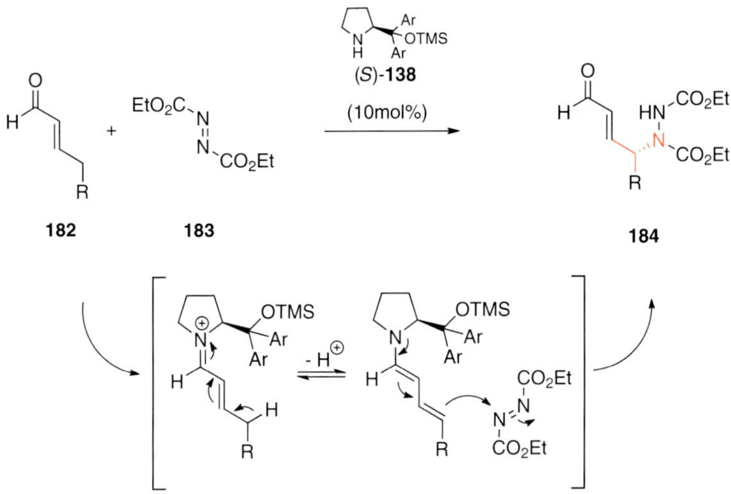

Scheme 43 *Jørgensen's* dienamine-catalyzed γ-functionalization protocol (*166*)

One of the first applications of dienamine catalysis in natural product synthesis was reported by *Hong* and co-workers, who described an efficient organocatalytic synthesis of (+)-palitantin (**185**) (*167*). Palitantin is a highly oxygenated polyketide-derived fungal metabolite isolated from *Penicillium palitans* and *Penicillium frequentans* (*168*, *169*) and some interesting synthesis strategies have been reported so far (*167*, *170*, *171*). *Hong's* synthesis was based on an enantioselective formal [4 + 2] self-dimerization of the α,β-unsaturated aldehyde **186**. The reaction proceeded well in the presence of a higher amount of (*S*)-**12** (50 mol%) to give

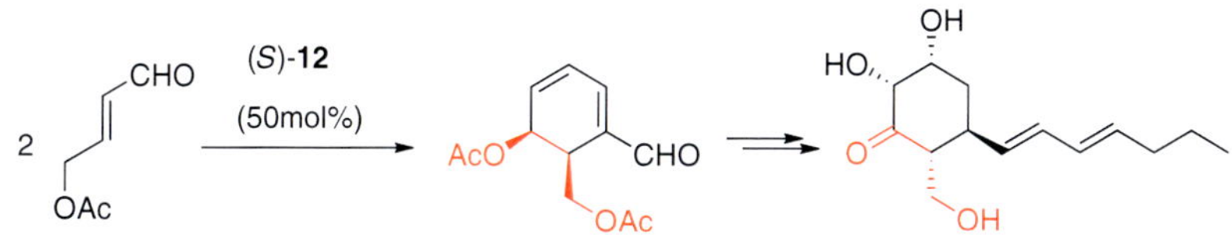

186 187 (77% *ee*, 80% *de*) (+)-Palitantin (185)

Scheme 44 Dienamine-catalyzed approach in the synthesis of (+)-palitantin (**185**)

the product **187** in reasonable selectivity. From a mechanistic point of view, the authors reasoned that the reaction more likely proceeded through a dienamine-catalyzed *Michael* reaction followed by an intramolecular *Mannich*-type addition rather than through a *Diels-Alder* reaction (*167*). Thus, and according to the mechanistic rationalization given by *Hong et al.*, this type of transformation can also be considered to be a domino or cascade process. Compound **187** was then converted successfully into (+)-palitantin (**185**) (Scheme 44). Further examples of dienamine catalysis in combination with other transformations or involved in cascade reactions will be given in Sect. 2.6.

Over the last few years the *Christmann* group has reported some interesting approaches towards the synthesis of mono- and bicyclic skeletons based on dienamine catalysis (*172*, *173*). As an example, (Scheme 45) compound (±)-**188**, a pungent constituent of black cardamom (*174*), was synthesized in a single transformation starting from the acyclic tethered α,β-unsaturated dialdehyde **189**. The bicyclic product **188** could be obtained in reasonable yield and good enantioselectivity through a formal [4 + 2] cycloaddition catalyzed by **138** in the presence

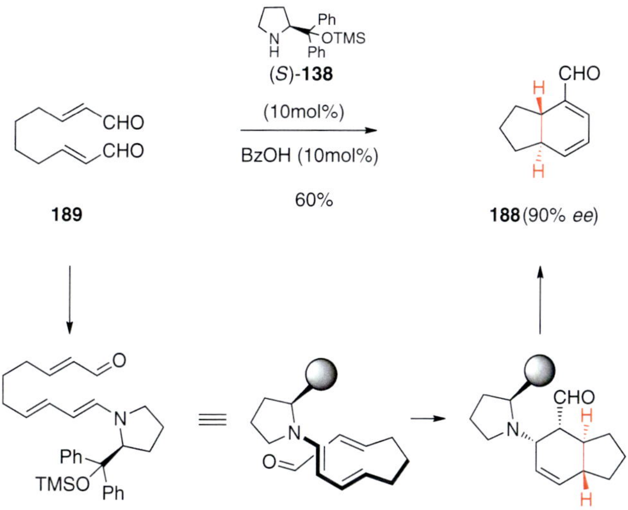

189 188 (90% *ee*)

190 191

Scheme 45 Dienamine-catalyzed synthesis of bicyclic scaffolds according to *Christmann et al.*

of small amounts of benzoic acid (BzOH). Mechanistically, this elegant transformation can be rationalized by the formation of the dienamine intermediate **190** first, which then undergoes a [4 + 2] cycloaddition to give **191** followed by a β-elimination to deliver the target compound **188** (*172*).

In 2009, the *Christmann* group reported the application of a dienamine intermediate for *Rauhut-Currier*-type reactions (*173*). In this case, the α-position of the dienamine acts as the nucleophile in an intramolecular cyclization reaction giving access to functionalized monocyclic compounds. The applicability of this strategy was illustrated in the synthesis of (*R*)-rotundial (**193**), a mosquito repellent from the leaves of *Vitex rotundifolia* (*175*). Hence, an organocatalytic *Rauhut-Currier*-type reaction of dialdehyde **192** catalyzed by 20 mol% (*S*)-**138** gave (*R*)-rotundial (**193**) directly in good enantioselectivity, albeit in only a moderate yield (Scheme 46).

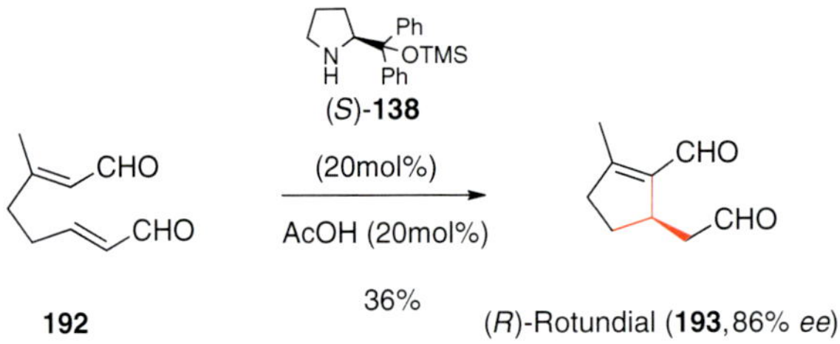

Scheme 46 Dienamine-catalyzed synthesis of (*R*)-rotundial (**193**)

2.6 Combined Enamine-Catalyzed Approaches and Cascade Reactions

The examples depicted so far have made use primarily of single organocatalytic transformations conducted typically quite early in the multi-step sequences applied towards the syntheses of complex natural products. In contrast, more and more reports describing organocatalytic cascade reactions or combined approaches using different organocatalytic key transformations to achieve a complex synthesis have been reported over the last several years (*30*, *32*, *176–178*). In this chapter, the application of combined enamine-catalyzed approaches for the syntheses of natural products will be described. Examples using different activation modes (*e.g.* enamine and iminium activation) will be discussed later.

One of the first examples of an enamine-catalyzed cascade reaction for the synthesis of a complex alkaloid was reported by *Itoh et al.* (*179*). Reaction of the dihydrocarboline **194** with enone **195** in the presence of (*S*)-**12** (7 days) gave the tetracycle **196** as a single diastereomer and in excellent enantiopurity (99%). This reaction can be described best as an enamine-catalyzed *Mannich-Michael* domino addition. Further manipulations then gave access to the indole alkaloid *ent*-dihydrocorynantheol (**197**) in an elegant and facile manner. As depicted in

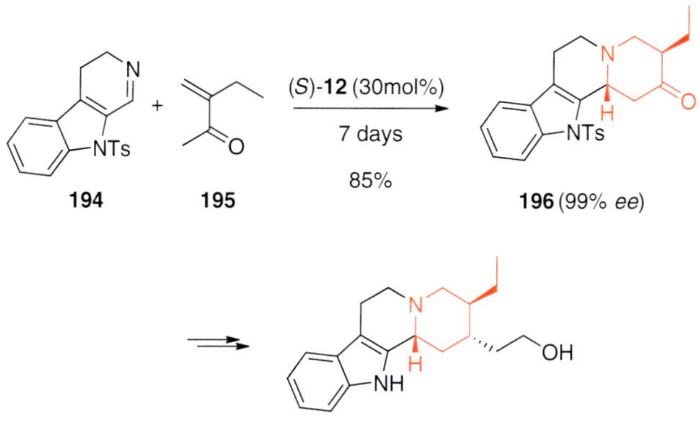

Scheme 47 Enamine-catalyzed *Mannich-Michael* addition in the synthesis of *ent*-dihydrocorynantheol (**197**)

Scheme 47, use of (*R*)-**12** as a catalyst in the first step should therefore give access to the natural product dihydrocorynantheol, a compound isolated from *Aspidosperma marcgravianum* showing activity against *Gram*-positive bacteria (*180*).

The combined use of an asymmetric α-oxygenation of an aldehyde followed by an olefination reaction and an organocatalytic conjugate addition was found to be a very useful approach for the syntheses of the complex natural products brasoside (**198**) and littoralisone (**199**) (*115*) (Scheme 48). Littoralisone (**199**) was found to be the active constituent of extracts of *Verbena littoralis* for increased nerve growth factor (NGF)-induced neurite outgrowth in PC12D cells (*181*) and is presumed to be derived biochemically from brasoside (**198**) (*182*, *183*). In 2005, *MacMillan* and *Mangion* reported the first total syntheses of **198** and **199** using three proline (**12**)-catalyzed transformations early in the synthesis. The *cis*-bicyclic skeleton was obtained successfully starting from the (−)-citronellol-derived aldehyde **200**. An asymmetric α-oxygenation catalyzed by (*R*)-**12** followed by a *Horner-Wadsworth-Emmons* (*HWE*)-olefination gave **201** initially. Redox-state manipulations then gave the dialdehyde **202**. This formyl-enal *Michael* acceptor was submitted further to a (*S*)-**12**-catalyzed intramolecular conjugate addition. The crucial part in this transformation was the choice of solvent. While $CHCl_3$ resulted in the formation of the *trans*-isomer, the use of DMSO gave the kinetic *cis*-product **203** in good yield and selectivity upon *in situ* O-acylation of the intermediate lactol. The iridoid **203** was then converted into the lactone **204** by standard methods. This key intermediate was used to accomplish the total syntheses of (−)-brasoside (**198**) and (−)-littoralisone (**199**) in a straightforward manner. Notably, the synthesis of **199** required the coupling partner **205** that was obtained easily using *MacMillan's* two-step carbohydrate protocol (Scheme 16) (*71*). Coupling of **204** and **205** followed by an impressive light-induced [2 + 2] cycloaddition and deprotection sequence then gave (−)-littoralisone (**199**) (*115*).

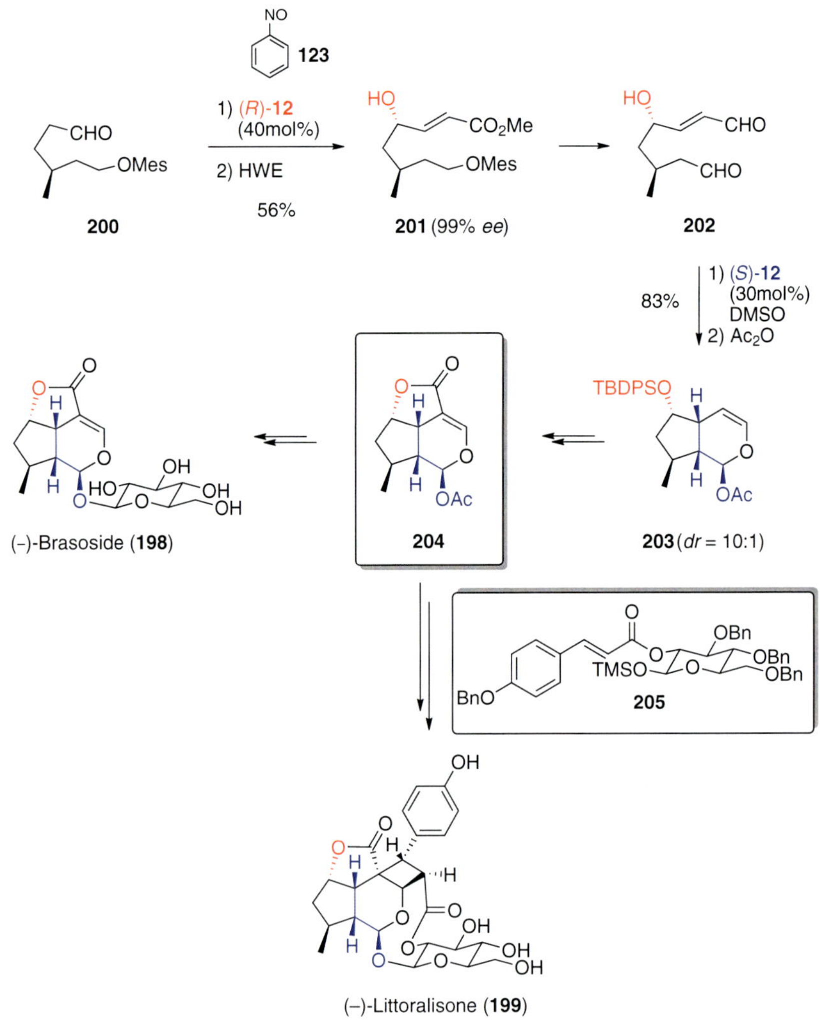

Scheme 48 *MacMillan's* enamine catalysis-based total syntheses of (−)-brasoside (**198**) and (−)-littoralisone (**199**)

Another excellent example from the *MacMillan* laboratory using enamine-catalyzed key-transformations to accomplish a complex total synthesis was the first total synthesis and structural revision of the cytotoxic marine macrolide callipeltoside C (**206**) (*184*). Callipeltosides A-C were isolated and characterized in 1996 and 1997 by *Minale et al.* They were found to be cytotoxic against the human bronchopulmonary NSCLC-N6 cell line (IC_{50} values ranging from 11.3 to 30.0 μg/cm^3) (*185*, *186*). *MacMillan's* synthesis of **206** takes advantage of three highly selective organocatalytic key transformations to obtain the target in 20 steps and a very

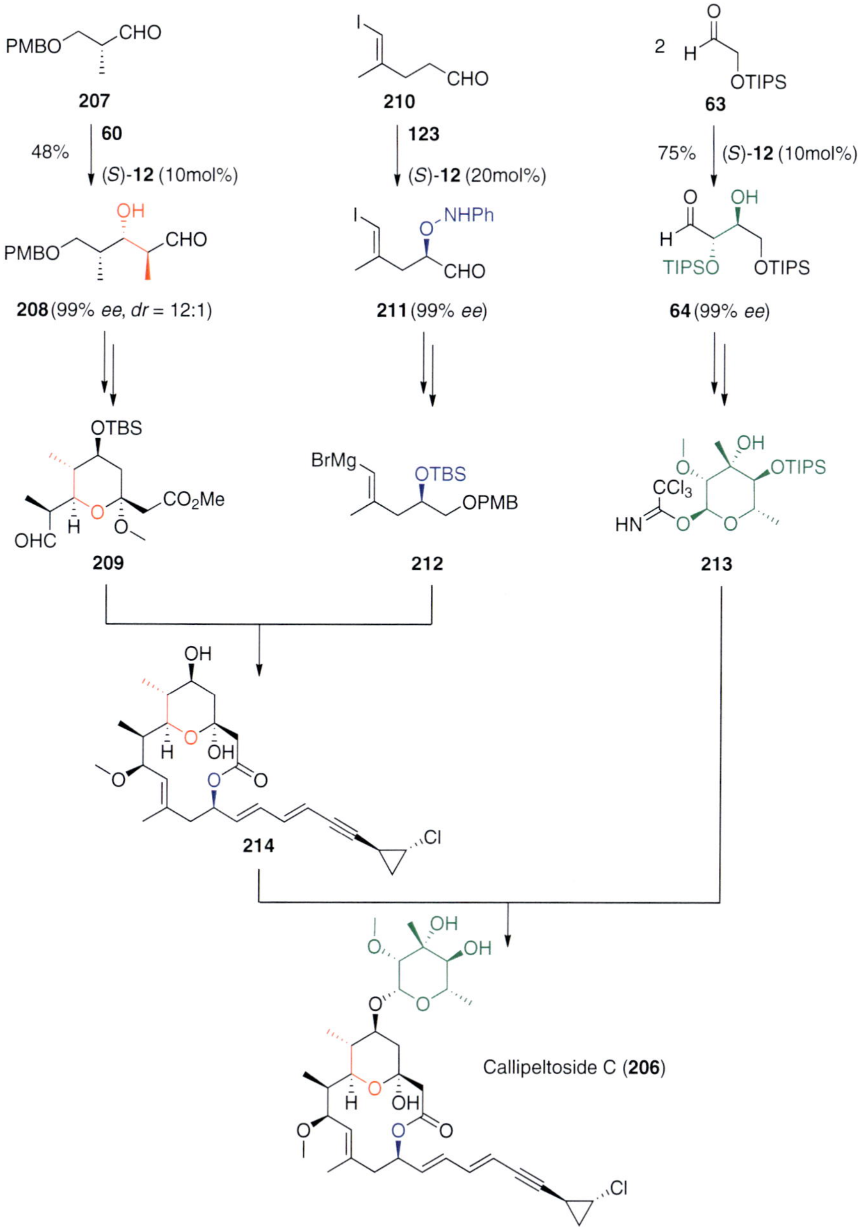

Scheme 49 Total synthesis and structural revision of callipeltoside C (**206**)

satisfactory overall yield of 11% (*184*). The (*S*)-**12**-catalyzed aldol reaction between **60** and **207** gave **208** in high selectivity. The aldol product **208** was progressed towards the functionalized tetrahydropyran **209**, a key intermediate in the overall assembly strategy. The second key fragment **212** was obtained after an enantioselective α-oxygenation of **210** to give **211** followed by standard transformations. The L-callipeltose-based third fragment **213** was synthesized successfully by the group's trademark carbohydrate protocol (Scheme 16) (*71*). Coupling of fragments **209** and **212** followed by further transformations furnished the aglycone **214**. Finally, the originally proposed structure of **206** suggested a glycosylation of **214** with the enantiomer of **213** (which was prepared in the same way as **213**, but using (*R*)-**12** as a catalyst). However, comparison of the spectroscopic data of the natural compound and the synthetic version revealed a significant difference. In contrast, coupling of **214** with **213** gave **206**, as shown in Scheme 49, and in full spectroscopic data accordance of the synthetic compound with the naturally isolated product (*184*).

Accordingly, the total synthesis of callipeltoside C (**206**) not only showed the high potential of organocatalysis in complex natural product synthesis, but it also pointed out the advisability of proving initially proposed structures of natural compounds by total synthesis.

Woggon and co-workers reported recently (Scheme 50) a highly diastereoselective domino aldol-oxa-*Michael* reaction of salicylaldehyde **215** and phytenal

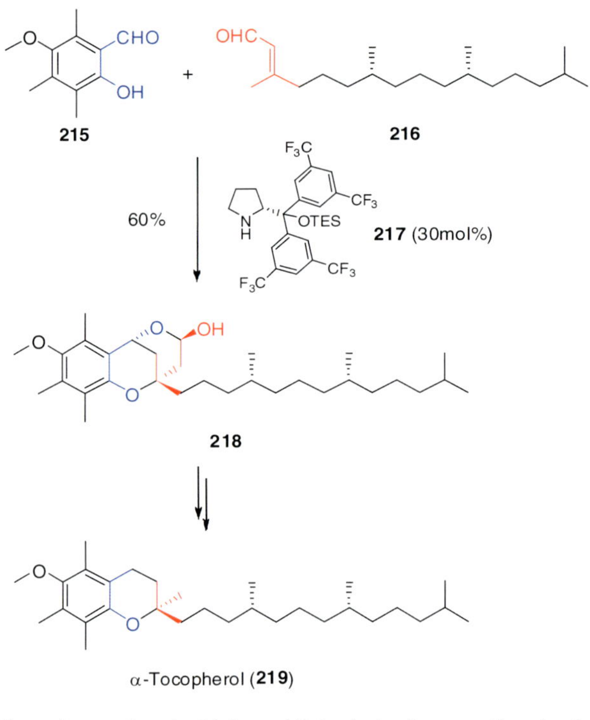

Scheme 50 Dienamine-catalyzed aldol-oxa-*Michael* domino reaction in the synthesis of α-tocopherol (**219**)

(**216**) in the presence of catalyst **217**, giving the hemiacetal **218** (*187*). Formation of this product can be explained by an initial dienamine-catalyzed aldol reaction of **216** to **215**, followed by an intramolecular oxa-*Michael* addition of the phenolic OH-group to the α,β-unsaturated iminium intermediate. Thus, from the mechanistic point of view, this domino reaction can be considered to make use of two different activation modes, namely, a dienamine activation first enabling the aldol reaction and an iminium activation then enabling the oxa-*Michael* addition (more distinct examples of iminium catalysis will be given in the next chapter). Finally, the thus obtained hemiacetal **218** then allowed the synthesis of α-tocopherol (**219**) in a direct and facile way (*187*).

2.7 Synopsis

Although enamine catalysis has only been attracting the attention of a broader audience for approximately the last 10 years, this activation mode has (together with iminium catalysis) contributed to possibly the most significant recent progress in the field of asymmetric organocatalysis. This methodology belongs to the most useful and broadly applicable current strategies to carry out a variety of different α-carbonyl reactions in a stereoselective fashion. As shown by previous examples, this methodology has quickly found its way into the toolbox of synthesis-oriented organic chemists. The following table gives a summary of enamine-catalyzed reactions presented in Section 2. (Table 1). As can be seen from this table, by far the most commonly employed catalyst included therein is proline (**12**).

Table 1 Enamine catalysis employed in complex natural product syntheses

Catalyst	Product	References
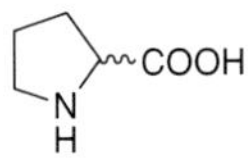 **12**	(*S*)-Ipsenol (**22**)	(*44*)
	(−)-6-Acetoxyhexadecanolide (**28**)	(*45*, *46*)
	Epothilone B (**29**)	(*47*)
	D-*arabino*-Phytosphingosine (**38**)	(*58*)
	L-*ribo*-Phytosphingosine (**39**)	(*58*)
	D-Psicose (**45**)	(*57*)
	D-KDG (**46**)	(*60*)
	Salinosporamide A (**56**)	(*68*)
	Prelactone B (**62**)	(*70*)
	O-Protected Hexoses (**66**, **67**, **68**)	(*71*)
	(+)-Spongistatin 1 (**69**)	(*78*)
	Trichostatin A (**74**)	(*79*)
	(+)-Cocaine (**92**)	(*86*)
	Quinine (**2**)	(*91*)
	Quinidine (**3**)	(*91*)
	(+)-Coniine (**105**)	(*94*, *95*)
	ent-Sedridine (**106**)	(*94*, *95*)
	Nikkomycin B (**107**)	(*96*)
	(+)-Polyoxamic acid (**112**)	(*97*)
	(+)-*epi*-Cytoxazone (**119**)	(*99*)
	Protected amino sugars (**122**)	(*101*)
	Brevicomin (**124**)	(*111*)
	RK-805 (**129**)	(*112*)
	(+)-Panepophenanthrin (**131**)	(*113*)
	Neosymbioimine (**134**)	(*116*)
	Disparlure (**135**)	(*120*)
	Piperazic acid (**140**)	(*128*)
	(+)-Palitantin (**185**)	(*167*)
	ent-Dihydrocorynantheol (**197**)	(*179*)
	(−)-Brasoside (**198**)	(*115*)
	(−)-Littoralisone (**199**)	(*115*)
	Callipeltoside C (**206**)	(*184*)
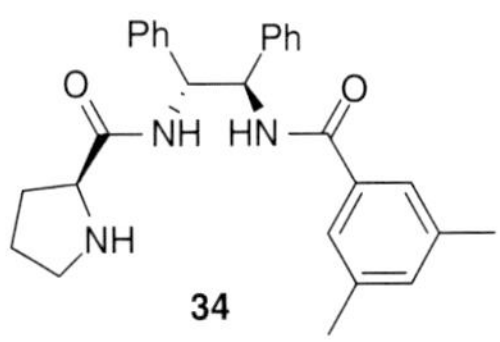 **34**	(*S*)-Convolutamydine A (**32**)	(*52*)
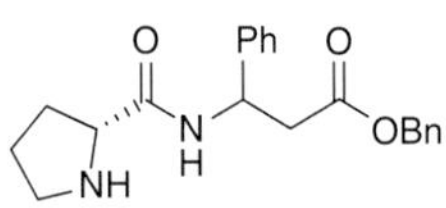 **35**	(*R*)-Convolutamydine A (**32**)	(*51*)

(continued)

Table 1 (continued)

Catalyst	Product	References
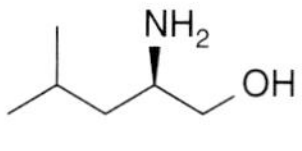 **36**	(*R*)-Convolutamydine A (**32**)	(*53*)
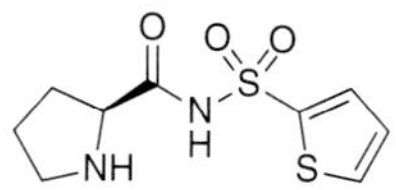 **37**	(*R*)-Convolutamydine A (**32**) (*R*)-Convolutamydine E (**77**) (*R*)-Convolutamydine B (**78**)	(*54*) (*80*, *81*) (*80*, *81*)
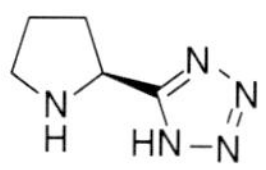 **54**	Serricornin (**51**)	(*62*)
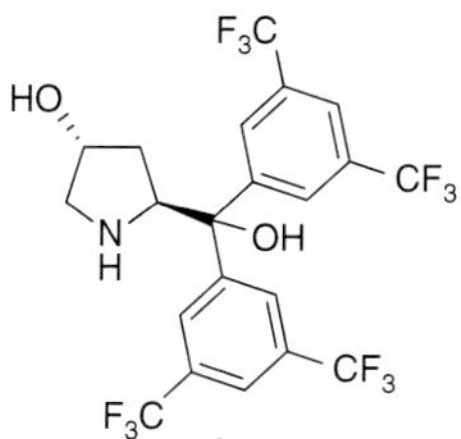 **81**	(*S*)-Convolutamydine E (**77**) CPC-1 (**86**) Madindoline A (**88**), B (**89**)	(*82*) (*83*) (*84*)
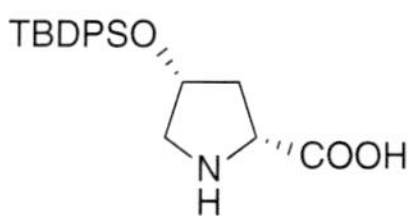 **95**	(+)-Juvabione (**96**)	(*89*)
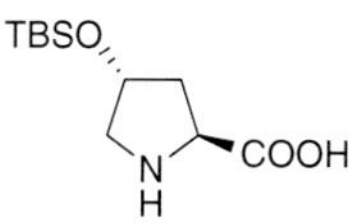 **121**	Protected amino sugars (**122**)	(*101*)

(continued)

Table 1 (continued)

Catalyst	Product	References
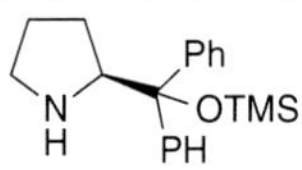 **138**	(−)-6-Acetoxyhexadecanolide (**28**)	(*121*)
	Cermizine C (**144**)	(*138*, *139*)
	Cermizine D (**145**)	(*138*, *139*)
	Senepodine G (**146**)	(*138*, *139*)
	Cernuine (**147**)	(*138*, *139*)
	(*R*)-Rotundial (**193**)	(*173*)
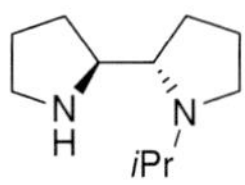 **152**	(−)-Botryodiplodin (**155**)	(*153*)
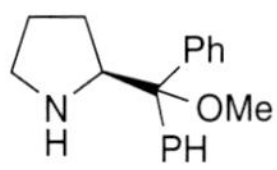 **159**	Biyouyanagin A (**156**)	(*155*)
	Eudesmane terpenes (**163–166**)	(*158*)
	(+)-Polyanthellin A (**170**)	(*160*)
	(−)-Bitungolide F (**172**)	(*161*)
	(−)-Clavukerin A (**176**)	(*162*)
	(−)-Isoclavukerin A (**177**)	(*162*)
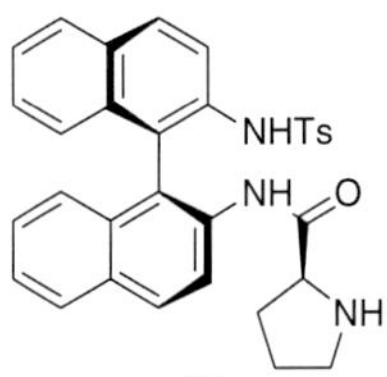 **180**	Anominine (**178**)	(*164*)
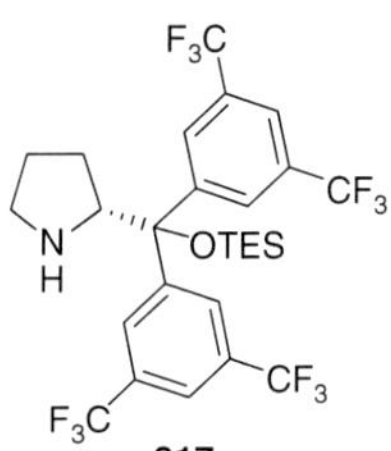 **217**	α-Tocopherol (**219**)	(*187*)

3 Iminium Catalysis

Besides the use of chiral amines to activate nucleophiles *via* enamine formation, their use as acceptor-activating catalysts *via* iminium formation has resulted in the development of numerous impressive applications over the last decade. The concept of iminium activation of conjugated enones has been postulated to play a pivotal role in the biogenesis of natural products as proposed by *Baldwin et al.* for the key *Diels–Alder* step in *e.g.* the biosynthesis of himgravine (*188–190*).

In 2000, *MacMillan's* seminal report on the use of chiral imidazolium salt catalysts for *Diels-Alder* reactions (Scheme 3) (*27*) introduced a highly useful and generalized strategy for asymmetric C–C bond-forming reactions by activating an α,β-unsaturated carbonyl compound with catalytic amounts of a chiral secondary amine (*21*, *22*, *24*, *26*). The conceptual breakthrough was the realization that the *in situ* iminium formation between the catalyst and the enone results in a significant LUMO-lowering of the electrophile by analogy with the LUMO-lowering effect of commonly employed *Lewis* acid catalysts (Scheme 51).

Moreover, in contrast to *Lewis* acids, the chiral information is much closer to the reactive site when a chiral amine forms the iminium intermediate, thus resulting in a significant discrimination of the two faces of the acceptor molecule. Therefore, it

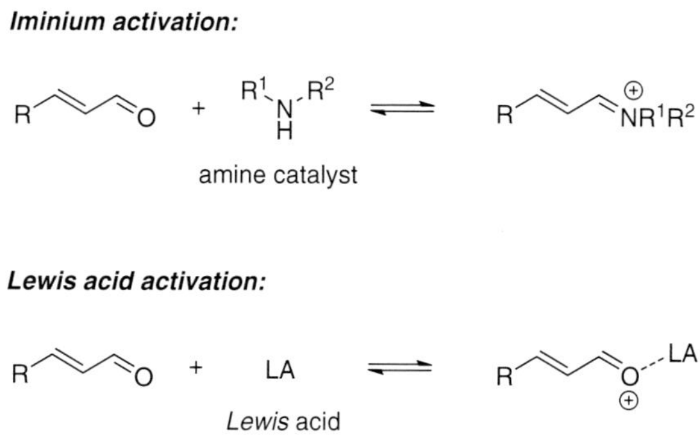

Scheme 51 LUMO-lowering activation of α,β-unsaturated carbonyl compounds by secondary amines and *Lewis* acids

M. Waser, *Asymmetric Organocatalysis in Natural Product Syntheses*,
Progress in the Chemistry of Organic Natural Products 96,
DOI 10.1007/978-3-7091-1163-5_3, © Springer-Verlag Wien 2012

can be said without exaggeration that the concept of iminium activation resulted in the development of some of the most useful organocatalytic transformations known today (*e.g.* cycloadditions or conjugate additions). Furthermore, these are often stereoselective reactions that are otherwise hard to achieve in such an easy and general way as with the use of a simple chiral amine. Accordingly, this novel strategy has set the stage for the development of some extraordinarily short and straightforward syntheses of complex (natural) products.

3.1 Pericyclic Reactions

Following the seminal report of *MacMillan et al.* in 2000 (*27*), several applications of iminium ion-activated *Diels-Alder* reactions in complex natural product syntheses have been reported, either carrying out the cyclization in an intra- or an intermolecular fashion.

3.1.1 *Intermolecular* Diels-Alder *Reactions*

One of the first applications of *MacMillan's Diels-Alder* strategy in a complex natural product synthesis was reported by *Kerr et al.* (Scheme 52) in their synthesis of (+)-hapalindole Q (**220**) in 2003 (*191*). The hapalindoles are a group of tri- and tetracyclic natural alkaloids that were first isolated by *Moore* and co-workers from

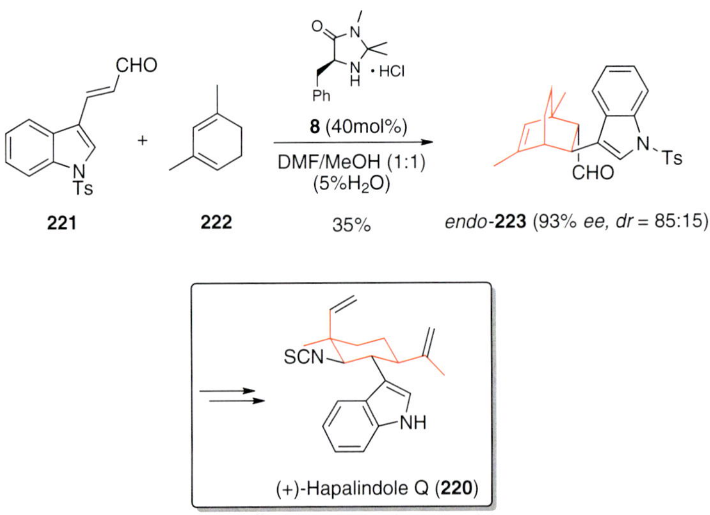

Scheme 52 "Organomediated" intermolecular *Diels-Alder* reaction in the total synthesis of (+)-hapalindole Q (**220**)

the terrestrial blue-green alga *Hapalosiphon fontinalis* after it was found that extracts of this alga exhibited antialgal and antimycotic activity (*192*, *193*). The key step in *Kerr's* synthesis was an intermolecular *Diels-Alder* reaction between enone **221** and the diene **222** catalyzed by *MacMillan's* catalyst **8**. Interestingly, in this special case the catalyst loading was higher than the actual yield, thus this reaction was considered not to be organocatalytic but organomediated. However, *Kerr et al.* also showed that the reaction could be carried out with catalytic amounts of **8**, albeit with lower selectivity (*191*).

The *MacMillan* group reported recently (Scheme 53) a nine-step synthesis of the complex *Strychnos* alkaloid (+)-minfiensine (**224**) starting from commercially available compounds (*194*). The key step in this elegant approach was an intermolecular highly enantioselective [4 + 2] cycloaddition between **225** and propynal

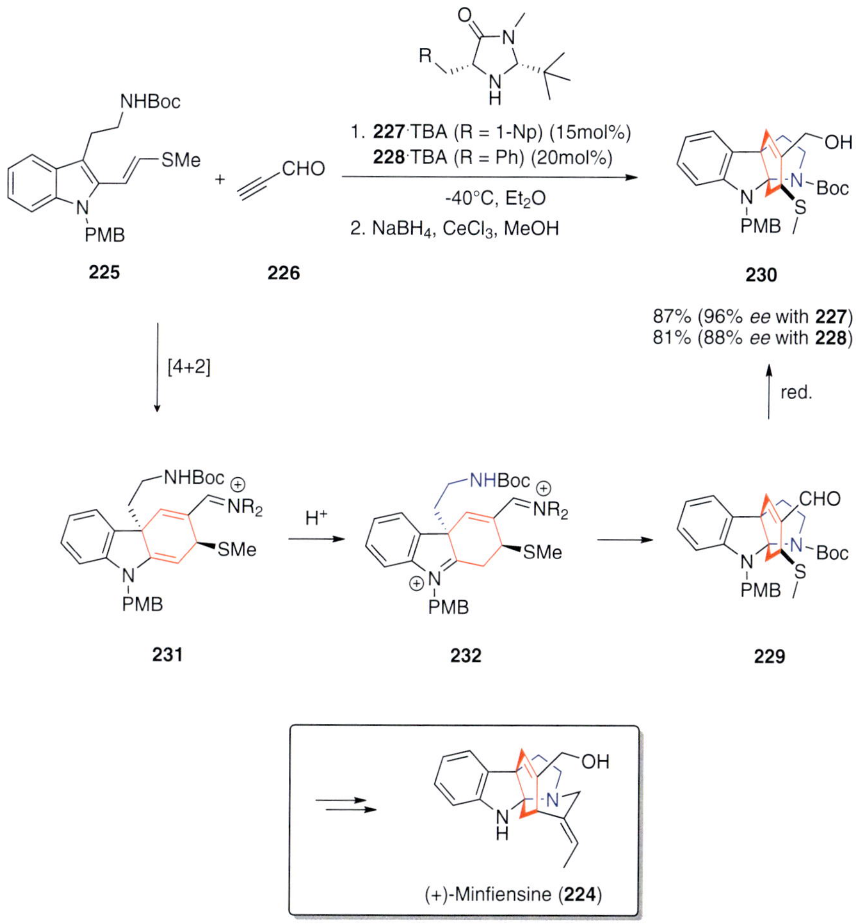

Scheme 53 Organocatalytic *Diels-Alder* cyclization cascade in the total synthesis of (+)-minfiensine (**224**)

(**226**) catalyzed by 15 mol% of the tribromoacetic acid (TBA) salt of imidazolidinone **227** (Scheme 53). The reaction gave directly the tetracyclic **229** which, upon *in situ* $NaBH_4$ reduction, gave the isolated alcohol **230**. This remarkable transformation can be rationalized by a cascade sequence involving an *endo*-selective [4 + 2] addition giving intermediate **231** first, followed by a protonation resulting in **232** and a final intramolecular amine cyclization to yield the product **229**. Of note, the naphthyl-substituted catalyst **227** was found to be superior with respect to yield and enantioselectivity compared to the corresponding phenyl-substituted **228**. Attempts to lower the catalyst amount to 5 mol% resulted in a slightly reduced yield of less than 80% but with similar enantioselectivity (94% *ee*). Further transformations to build up the final pentacyclic core then gave access to (+)-minfiensine (**224**) in 21% overall yield (*194*).

3.1.2 Intramolecular Diels-Alder *Reactions*

Besides intermolecular organocatalytic *Diels-Alder* approaches, intramolecular approaches also have been employed successfully on several occasions (*195–198*). *MacMillan et al.* were among the first to use such a strategy in natural product syntheses (*195*). As depicted in Scheme 54, treatment of the unsaturated aldehyde **233** with 20 mol% of the trifluoromethanesulfonate salt of catalyst **228** gave the bicyclic **234** in good yield and high selectivity. The aldehyde **234** was then converted into solanapyrone D (**235**) in five steps. Solanapyrone D (**235**) is a phytotoxic polyketide isolated from the fungus *Altenaria solani* (*199, 200*). *MacMillan's* synthesis of **235** is especially impressive if one keeps in mind that the first total synthesis of **235** in 2002 by *Hagiwara et al.* required 19 steps to obtain the target successfully (*201*), whereas the organocatalytic approach depicted in Scheme 54 consists of only six steps from the commercially available starting material **233** (*195*).

A rather similar approach was recently reported by *Koskinen et al.* who used an intramolecular iminium-catalyzed *Diels-Alder* reaction to build up the bicyclo [4.3.0]nonane skeleton of amaminol A (**236**) and amaminol B (**237**) (*196, 197*).

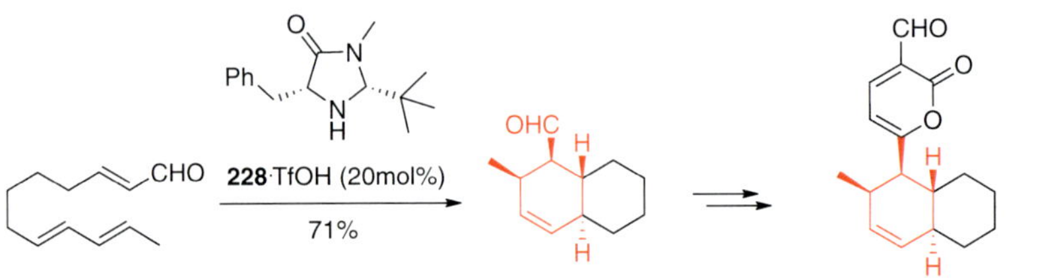

233 **234** (90% *ee*, *dr* > 20:1) Solanapyrone D (**235**)

Scheme 54 Intramolecular *Diels-Alder* reaction in the total synthesis of solanapyrone D (**235**)

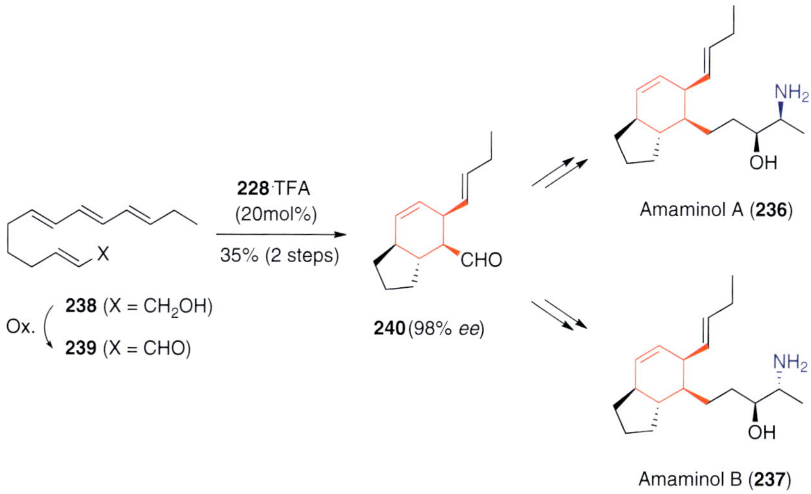

Scheme 55 Intramolecular *Diels-Alder* reaction in the total syntheses of amaminol A (**236**) and amaminol B (**237**)

Amaminols A (**236**) and B (**237**) are two bicyclic amino alcohols isolated from an unidentified tunicate from the Amami islands exhibiting moderate cytotoxicity against P388 murine leukemia cells (*202*). The required key intermediate **240** was obtained in moderate yield and excellent stereoselectivity was observed by treatment of the *in situ*-prepared aldehyde **239** with catalytic amounts of the TFA salt of **228**. The bicyclo[4.3.0]nonane **240** was then successfully converted further either into amaminol A (**236**) or amaminol B (**237**) (Scheme 55).

3.2 Conjugate Additions

Although *Diels-Alder* reactions provided the first examples for the great potential of asymmetric iminium catalysis, it must be pointed out that by far the most applications of this type of catalysis in natural product syntheses have been reported for conjugate additions of different nucleophiles to iminium-activated α,β-unsaturated acceptor molecules. The following sections will give an overview based on the type of nucleophiles employed in such transformations.

3.2.1 Conjugated Transfer Hydrogenations

The iminium-catalyzed reduction of α,β-unsaturated carbonyl compounds using *Hantzsch* dihydropyridines as the hydride source was reported independently by the *List* and *MacMillan* groups at the end of 2004 and the beginning of 2005 (*203–205*).

Mimicking Nature's reducing agent NADH (nicotinamide adenine dinucleotide), the *Hantzsch* dihydropyridines were found to be very useful transfer hydrogenation agents enabling highly stereoselective reductions of α,β-unsaturated electrophiles in combination with a chiral catalyst.

Besides the utilization of chiral secondary amines to achieve a LUMO-lowering activation as well as face discrimination, the use of achiral secondary amines in combination with a chiral counterion also proved to be highly promising for such transformations. This strategy resembles the use of achiral metal catalysts in combination with a chiral ligand to achieve a stereoselective transformation (*206–208*). It is due to *Benjamin List* that the elegant concept of asymmetric counteranion-directed catalysis (ACDC) has found widespread applications in organocatalysis at the present time (*209–212*).

Using catalytic amounts of the morpholine salt of a chiral phosphoric acid such as compound **241** and *Hantzsch* ester **242** as the hydride source, *List et al.* were able to achieve highly selective reductions of a broad variety of α,β-unsaturated carbonyl compounds like farnesal (**243**) as demonstrated in the enantioselective synthesis of the bee pheromone (*R*)-**244** (*210*) (Scheme 56). Notably, this method was found to be superior when compared to the use of chiral amine-based catalysts with respect to enantioselectivity in several examples employing sterically unhindered aliphatic aldehydes (*209*).

A chiral amine-catalyzed transfer hydrogenation was recently employed successfully by *Lear* and co-workers to obtain a key intermediate for the total synthesis of (−)-platensimycin (**245**) (*213*). (−)-Platensimycin (**245**) was identified from *Streptomyces platensis* in 2006 (*214*, *215*). Due to its impressive antibacterial properties, it has generated considerable interest within the scientific and medical communities as a potential powerful new therapy against drug-resistant bacteria (*214*, *215*). Accordingly, several (formal) total synthesis approaches have been

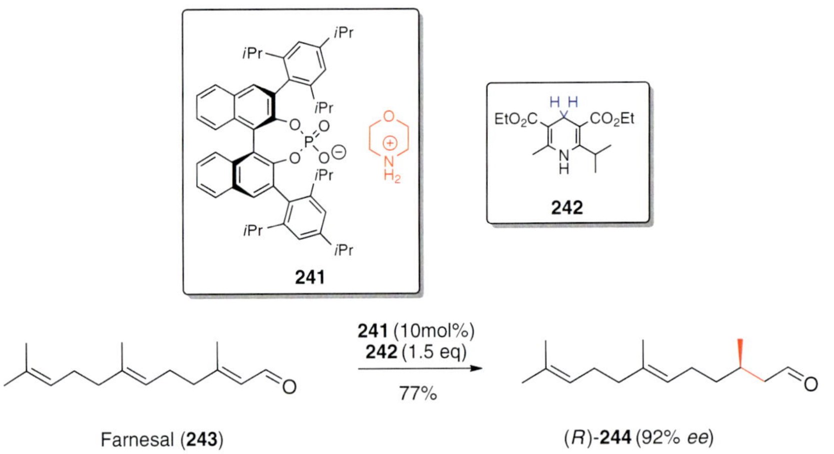

Scheme 56 Asymmetric counteranion-directed transfer hydrogenation in the synthesis of the bee pheromone (*R*)-**244**

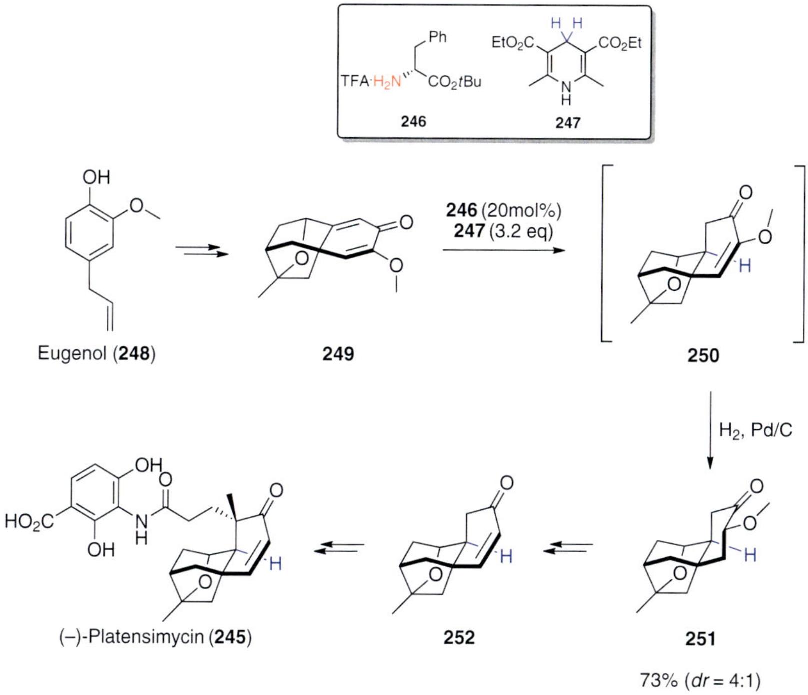

Scheme 57 Diastereoselective transfer hydrogenation in the formal total synthesis of (−)-platensimycin (**245**)

reported over the last few years (*216–220*). In their formal approach towards **245**, *Lear et al.* employed the D-phenylalanine-derived catalyst **246** in combination with the *Hantzsch* ester **247** to achieve the chemo- and stereocontrolled conjugate reduction of olefin **249** (synthesized in a very impressive fashion from eugenol (**248**) (*213*)) to furnish the targeted **250**. It is worth mentioning that *Lear et al.* tested a variety of different hydrogenation approaches (*e.g.* different metal-catalyzed ones or other secondary amine catalysts) to achieve this transformation, but none of these gave a reasonable stereocontrol, thus making the organocatalytic approach the method of choice for this step. The transfer hydrogenation product mixture was then directly submitted to a Pd/C-mediated heterogeneous hydrogenation to obtain a separable mixture of the diastereomers of **251** (Scheme 57) (*213*). Compound **251** could then be transformed successfully into *Nicolaou's* tetracyclic enone **252**, a well-documented key intermediate in the synthesis of (−)-platensimycin (**245**) (*216, 217*).

The imidazoline salt **253** was reported recently by *Willis et al.* to be the catalyst of choice in an iminium-catalyzed transfer hydrogenation (using ester **247** as a hydride donor) of an (*E*)/(*Z*)-mixture of the α,β-unsaturated aldehyde **254** to furnish the chiral aldehyde **255** in good yield and enantioselectivity (*221*). With compound

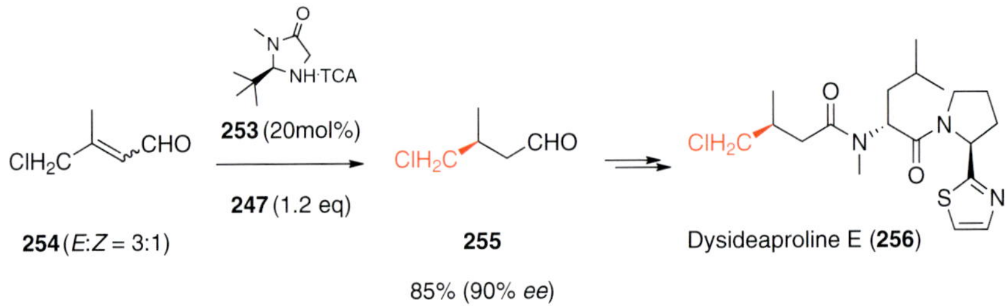

Scheme 58 Enantioselective transfer hydrogenation in the synthesis of (−)-dysideaproline E (**256**)

255 in hand, the total synthesis of dysideaproline E (**256**) could be accomplished in a short period of time and in a straightforward manner (Scheme 58). Dysideaproline E (**256**) belongs to a family of chlorinated natural products isolated from extracts of a *Dysidea* sp. from the Philippines (*222*).

3.2.2 Carbon Nucleophiles in Michael-Type Reactions

Carbon-carbon bond forming reactions between carbanionic nucleophiles like enolates or deprotonated nitroalkanes and electron deficient alkenes and alkynes belong to the oldest and most versatile transformations known today (*223–229*). Moreover, stereoselective variants have proven to possess an enormous potential in the syntheses of complex molecules as already exemplified in Sect. 2.4. Whereas the applications depicted in this previous section utilized nucleophiles activated by enamine formation with a chiral secondary amine catalyst to achieve these highly selective C–C bond formations, the present discussion will focus on the addition of carbon nucleophiles to iminium-activated *Michael* acceptors. Herein "traditional" *Michael* additions using *e.g.* enolate nucleophiles will be described whereas the use of aromatic *Michael* donors with iminium-activated acceptors in *Friedel-Crafts* type reactions will be discussed separately subsequently.

1,3-Dicarbonyl compounds are among the most prominent nucleophiles used in *Michael* reactions (*230–232*). *Jørgensen et al.* applied an enantioselective iminium-catalyzed *Michael* addition for the successful synthesis of the anticoagulant warfarin (**257**). Warfarin (or coumadin) (**257**) itself is not a natural product but a synthetic derivative of dicoumarol (**258**), a mycotoxin anticoagulant discovered in spoiled sweet clover (*233*). Warfarin is usually prescribed as a racemate, despite the fact that both enantiomers have different activities and different half-lives (*234*, *235*). Although it is not totally within the focus of this review to cover natural product-derived synthetic drugs, this example is definitely worth mentioning, since the organocatalytic approach of *Jørgensen et al.* gave access to (*R*)-**257** in a highly selective and impressively simple manner. This strategy is of major interest, as the

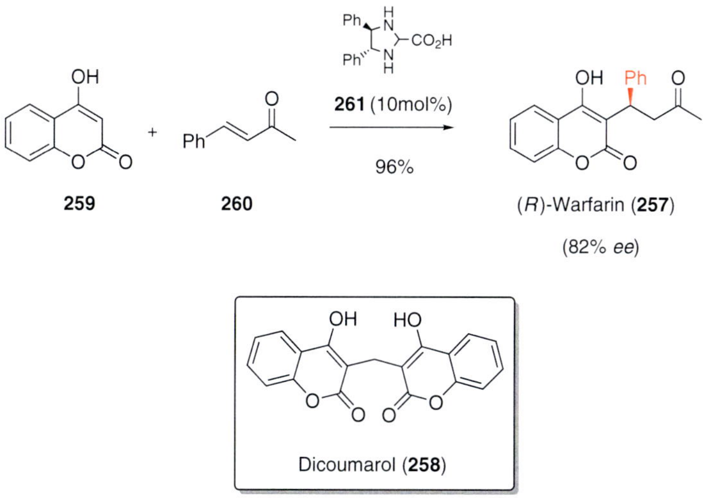

Scheme 59 Enantioselective *Michael* addition in the synthesis of the synthetic dicoumarol (**258**) analogue (*R*)-warfarin (**257**)

application of enantiopure warfarin would give access to a much better control in therapy. As depicted in Scheme 59, the targeted introduction of the stereogenic center could easily be achieved by reacting the hydroxycoumarin **259** with the benzylidenacetone **260** in the presence of 10 mol% of catalyst **261** (*230*). Similar strategies have also been applied by the *Jørgensen* (*231*) and the *Rios* (*232*) groups in their (formal) total syntheses of the antidepressant drug paroxetine.

The reaction of silyl-enol ethers with carbonyl compounds, also known as the *Mukaiyama*-aldol reaction (*236, 237*), represents one of the most useful methods for the (stereoselective) construction of carbon-carbon bonds (*238–240*). As an example of special interest, the γ-butenolide moiety is present in over 13,000 natural products and the coupling of silyloxy furans and aldehydes using chiral *Lewis* acids is one of the most versatile strategies for butenolide syntheses (*241–243*).

While *Lewis* acids normally enable 1,2-additions of silyloxy furans to α,β-unsaturated aldehydes, the *MacMillan* group discovered that iminium catalysis favors 1,4-additions, thus overcoming the deficiency of normal *Lewis* acids in *Mukaiyama-Michael* additions (*244*). The high potential of this protocol was demonstrated impressively in a short synthesis of the *Penicillium spiculisporum* fermentation product spiculisporic acid (**262**). As shown in Scheme 60, vinylogous addition of the silyloxy furan **263** to the acceptor **264** catalyzed by **228** furnished the key intermediate **265** in good enantio- and diastereoselectivity. It is worth noting that the TfOH salt of the catalyst in combination with methyl ester **266** as a *Michael* acceptor gave the *anti*-diastereomer **267** exclusively. This compound was then used for the synthesis of 5-*epi*-spiculisporic acid (**268**) (Scheme 60) (*244*).

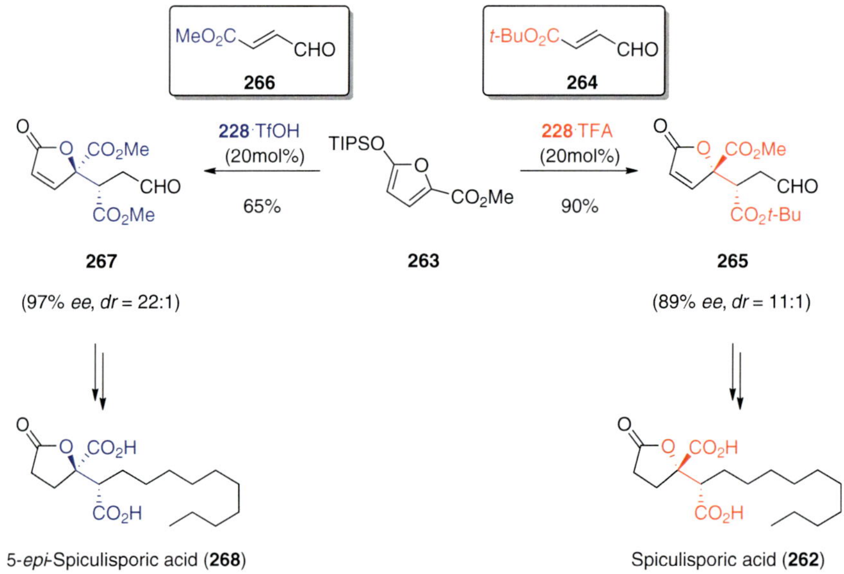

Scheme 60 Iminium-catalyzed *Mukaiyama-Michael* addition in the synthesis of spiculisporic acid (**262**)

A similar strategy was used by *Robichaud et al.* early in their formal total synthesis of (+)-compactin (mevastatin) (**269**) (*245*). Compactin is a potent inhibitor of HMG-CoA reductase and is a cholesterol-lowering drug that was isolated in the 1970s by *Endo et al.* from *Penicillium citrinum* (*246–248*). In their synthesis strategy towards **269**, *Robichaud et al.* employed a *MacMillan*-type conjugate addition of the furan **270** to the TMS-aldehyde **271** using the (*S*,*S*)-enantiomer of catalyst **228**. This furnished the intermediate aldehyde **272** in high yield and good selectivity (Scheme 61) (*245*).

As already mentioned earlier (*e.g.* Scheme 35), *Lycopodium* alkaloids have always been interesting targets in natural product syntheses (*129–133*). In 2007, the group of *Toste* reported a 13-step total synthesis of (+)-fawcettimine (**273**) with

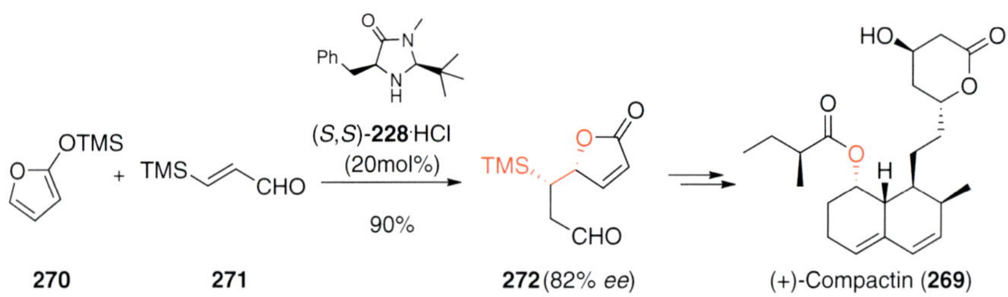

Scheme 61 Iminium-catalyzed *Mukaiyama-Michael* addition in the formal total synthesis of (+)-compactin (**269**)

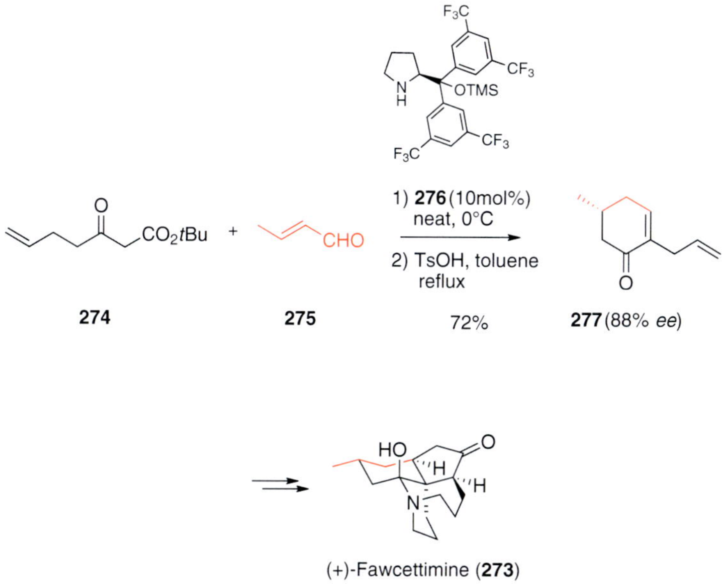

Scheme 62 Organocatalytic 1,4-addition in *Toste's* total synthesis of (+)-fawcettimine (**273**)

an organocatalytic 1,4-addition as one of the key steps (*133*). The fawcettimine class of *Lycopodium* alkaloids consists of over 60 natural compounds (*129*) and **273** was the first member to be isolated in 1959 by *Burnell* from a plant collected in the Blue Mountain Range of Jamaica (*249*). *Toste's* reaction sequence commenced with a vinylogous addition of ketoester **274** to crotonaldehyde (**275**) catalyzed by the secondary amine **276** (Scheme 62). The reaction furnished directly the cyclic **277** in good yield and enantioselectivity. Formation of **277** can be explained by a 1,4-addition first, followed by an intramolecular aldol condensation, and a final decarboxylation of the ester group. Intermediate **277** was then converted successfully into (+)-fawcettimine (**273**) in a series of highly efficient transformations. This is an example where the later synthesis steps are highly innovative and worth examining more closely. Therefore, the interested reader may be referred to the detailed discussion in the original work of *Toste et al.* (*133*).

Organocatalytic *Michael* additions using nitromethane (**278**) as a nucleophile have attracted considerable interest over the last years, resulting in the successful syntheses of natural and non-natural targets (*250–252*). Successful examples for the syntheses of designer drugs include the synthesis of the type IV phosphodiesterase inhibitor (*S*)-rolipram (*250*) and the GABA receptor agonist baclofen (*251*).

The cytotoxic alkaloid 7-deoxy-*trans*-dihydronarciclasine (**279**) was isolated from the bulbs of *Hymenocallis littoralis*, *Hymenocallis caribea*, and *Hymenocallis latifolia* by *Pettit et al.* in 1993 (*253*). Recently, *Kadas* and co-workers employed an iminium-catalyzed *Michael* addition between **278** and acceptor **280** to obtain the

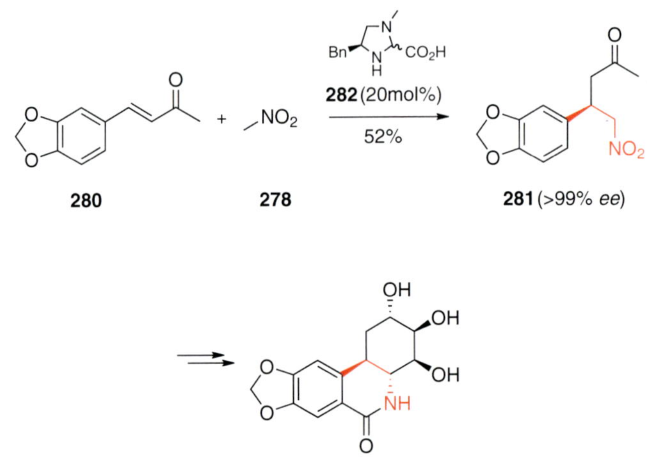

ent-(−)-7-Deoxy-*trans*-dihydronarciclasine (**279**)

Scheme 63 Synthesis of *ent*-(−)-7-deoxy-*trans*-dihydronarciclasine (**279**)

enantiopure **281** in reasonable yield. Compound **281** then served as the key synthon for the successful total synthesis of *ent*-7-deoxy-*trans*-dihydronarciclasine (**279**) (Scheme 63) (*252*).

3.2.3 Friedel-Crafts-*Type Reactions (Aromatic* Michael *Donors)*

Iminium catalysis was found to be also highly useful for the stereoselective functionalization of aromatic compounds *via Friedel-Crafts*-type alkylations. Pioneering work in this field came from the *MacMillan* group, who developed a highly enantioselective protocol for the reaction between heteroaromatic or electron-rich aromatic substrates and various *Michael* acceptors (*254–256*). One of the first applications of this strategy in natural product synthesis was reported in 2004 by *MacMillan et al.* employing it as a key step in their strategy towards the pyrroloindoline alkaloid (−)-flustramine B (**283**) (*257*). The flustramines are a family of marine alkaloids isolated from the bryozoan *Flusta foliacea* (*258–260*). Reacting the tryptamine derivative **284** with acrolein (**285**) in the presence of catalytic amounts of (*S*,*S*)-**228**.TsOH resulted in the formation of compound **286** in 90% *ee* and 78% yield. Formation of **286** can be explained by the *in situ* generation of *Friedel-Crafts* intermediate **287** initially, followed by an intramolecular cyclization to give **286** (Scheme 64). The natural product flustramine B (**283**) was then obtained successfully by standard transformations (*257*).

The *Banwell* group employed an organocatalytic intramolecular *Friedel-Crafts* cyclization to synthesize the alkaloids (−)-rhazinal (**288**), (−)-rhazinilam (**289**), (−)-leuconolam (**290**), and (+)-*epi*-leuconolam (**291**) (*261*). The spindle toxin (−)-

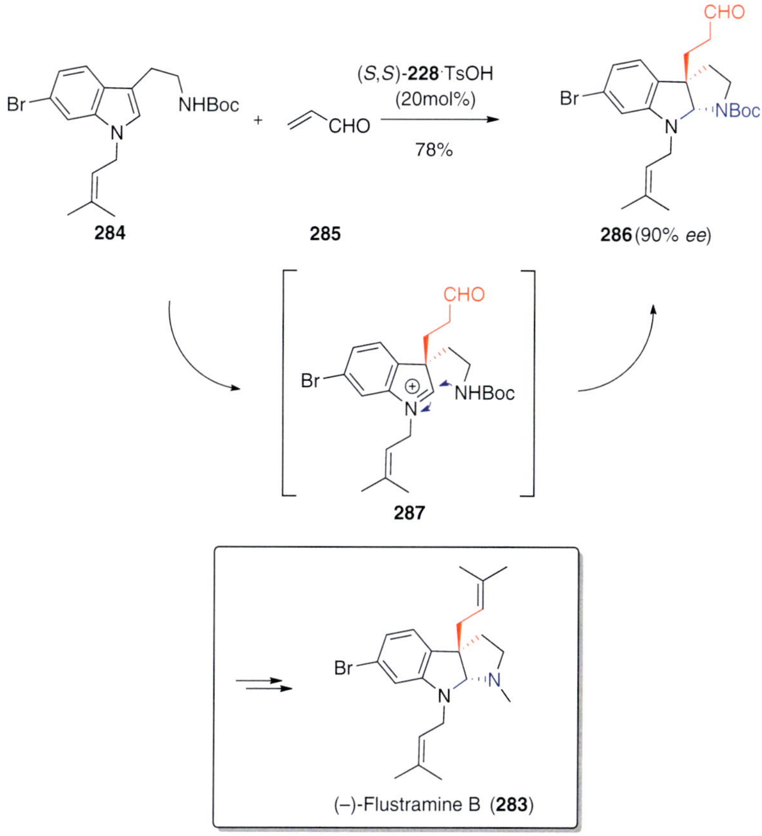

Scheme 64 Asymmetric *Friedel-Crafts* alkylation-cyclization cascade in the synthesis of flustramine B (**283**)

rhazinal (**288**) was isolated from the stem extracts of a Malaysian *Kopsia* species in 1998 (*262*). Like its congener rhazinilam (**289**), its potential to interfere with tubulin polymerization dynamics makes **288** a promising anticancer lead, which renders these compounds interesting targets for total syntheses and structural modifications (*263–266*).

Banwell et al. synthesized these alkaloids by carrying out an intramolecular cyclization of **292** using the *MacMillan* catalyst (**8**), which gave them the rather unstable key intermediate **293** in good yield and reasonable enantioselectivity. Compound **293** was then transformed into **288**, which could be converted further into the other natural analogues **289–291** (Scheme 65) (*261*).

Besides heteroaromatic compounds also electron-rich aromatic compounds like aniline or naphthol derivatives have been employed in organocatalytic *Friedel-Crafts* reactions (*267*, *268*). *Kim et al.* demonstrated this in the synthesis of (+)-curcuphenol (**294**) (*267*), a bioactive sesquiterpene phenol isolated from the marine

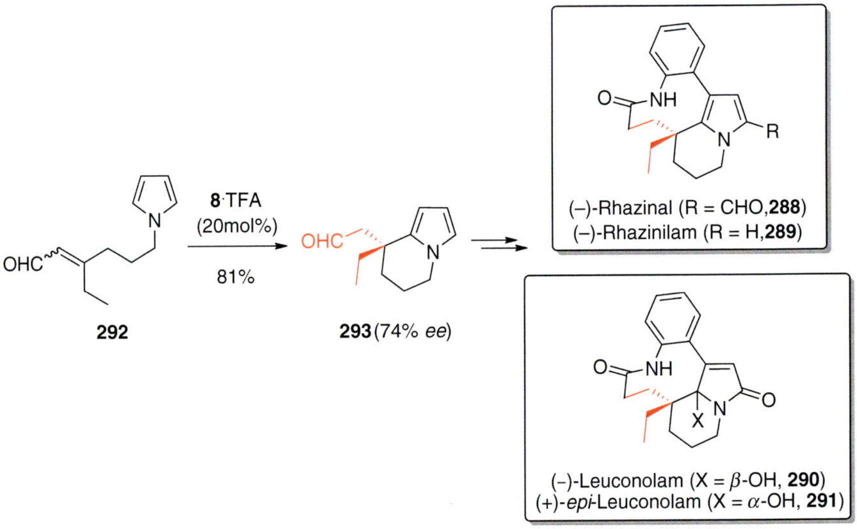

Scheme 65 Asymmetric *Friedel-Crafts*-type cyclization in the syntheses of the alkaloids (−)-rhazinal (**288**), (−)-rhazinilam (**289**), (−)-leuconolam (**290**), and (+)-*epi*-leuconolam (**291**)

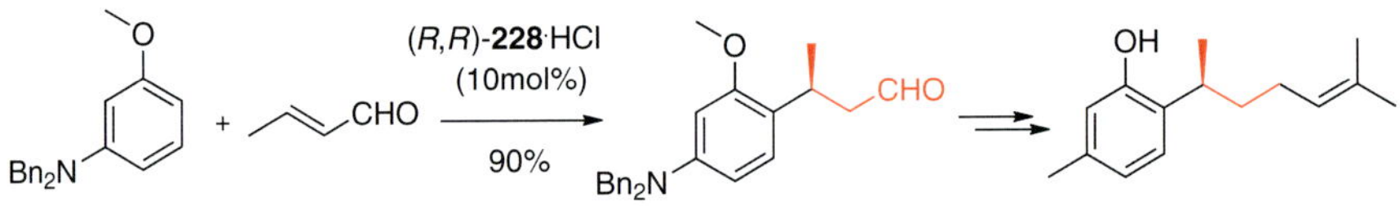

Scheme 66 *Friedel-Crafts* reaction in the synthesis of (+)-curcuphenol (**294**)

sponges *Didiscus flavus* and *Epipolasis species*. (+)-Curcuphenol (**294**) displays antifungal activity against *Candida albicans*, as well as cytotoxic activity against human cancer cell lines, antimalarial activity, and inhibition of ATPase (*269*, *270*). As depicted in Scheme 66, reaction of **295** with acceptor **275** in the presence of catalytic amounts of (*R*,*R*)-**228** gave aldehyde **296** in a good yield and with high enantioselectivity. The successful synthesis of (+)-**294** could then be accomplished in seven more steps (*267*).

Very recently, *MacMillan et al.* applied an intermolecular combined *Friedel-Crafts*-type conjugate addition/cyclization procedure as a key step in the total synthesis of diazonamide A (**297**) (*271*). Diazonamides are secondary metabolites isolated from the marine ascidian *Diazona* sp. (*272*, *273*). Diazonamide A (**297**) was found to be a potent antimitotic member of this structurally unique family, exhibiting low nanomolar GI_{50} values towards different human cancer cell lines (*272*, *274*). The unique structure of two 12-membered macrocycles that are conjoined through a triaryl-substituted quaternary stereogenic center (C–10) that is embedded in a furanoindoline core makes this compound a very interesting and

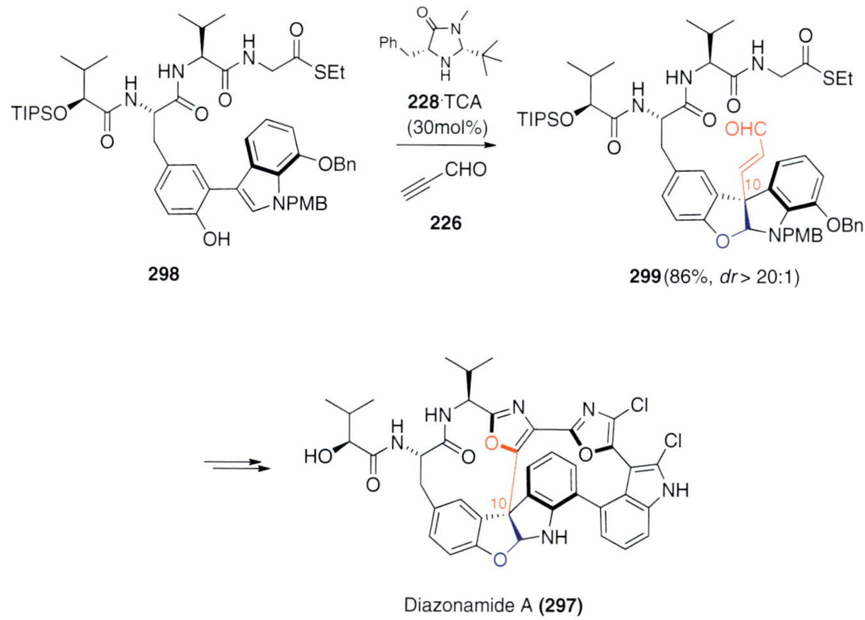

Scheme 67 *MacMillan's* combined addition/cyclization strategy for the synthesis of diazonamide A (**297**)

especially challenging target for total synthesis (*275–277*). *MacMillan's* strategy to introduce the targeted configuration of C-10 made use of an iminium-catalyzed reaction between the indole derivative **298** and aldehyde **226**. Mechanistically, the formation of the furanoindoline core of product **299** can be rationalized as a *Friedel-Crafts*-type conjugate addition of **298** to acceptor **226** initially, followed by an immediate nucleophilic addition of the phenol group to the *Friedel-Crafts* intermediate, by analogy to the sequence depicted earlier in the synthesis of flustramine B (**283**) (see Scheme 64). Interestingly, the diastereoselectivity of this transformation was influenced highly by the choice of solvent. Whereas a dichloromethane/methanol mixture gave a ratio of about 4:1, a ternary system of toluene, chloroform, and methanol resulted in an improved ratio of >20:1 (Scheme 67). Also, the use of a racemic catalyst resulted in a 1:1 mixture of diastereomers, thus illustrating the full stereocontrol of the catalyst. The target diazonamide A (**297**) was then obtained successfully in a series of further high yielding and efficient steps (*271*), making it another impressive example of the high potential of asymmetric organocatalysis in the total synthesis of complex natural products.

The application of iminium catalysis in *Friedel-Crafts* alkylations usually requires electron-rich aromatic partners. The *MacMillan* group investigated the possibility of overcoming this limitation of organocatalytic *Friedel-Crafts*-type alkylations using vinyl- or heteroaryl trifluoroborate salts as π-nucleophiles (*278*). In addition, the introduction of the trifluoroborate activation group increased significantly the scope of this reaction to a broad range of electron-neutral π-nucleophiles.

An impressive example of the high versatility of this elegant strategy in natural product synthesis was reported recently by the same group through their successful application of this method for the three-step total synthesis of (+)-frondosin B (**300**) (*279*). Frondosin B (**300**) is a marine sesquiterpene that was isolated initially from the marine sponge *Dysidea frondosa* in 1997 (*280*). The frondosin family has demonstrated inhibitory activity for the interleukin-8 cytotokine and for protein kinase C as well as showing potential anti-HIV properties (*281*). Due to this high possibility for medical applications, several groups have focused on the total synthesis of **300** or other members of this family (*279*, *282–285*). While most of this approaches required more than ten steps to give the natural product in lower overall yields, *MacMillan's* organocatalytic approach gave access to frondosin B (**300**) in three steps starting from commercially available starting materials with an overall yield of 50%.

Introduction of the tertiary stereogenic center could be achieved by reacting the commercial boronic acid **301** with the *Michael* acceptor **275** in the presence of one equivalent of HF to generate the activated boronate species *in situ*. The catalyst of choice in this approach was the tryptophan-derived imidazolidinone **302**. This strategy gave the aldehyde **303** in good yield and high enantioselectivity. Two further high yielding steps then gave (+)-frondosin B (**300**) in a very short and efficient procedure (*279*) (Scheme 68).

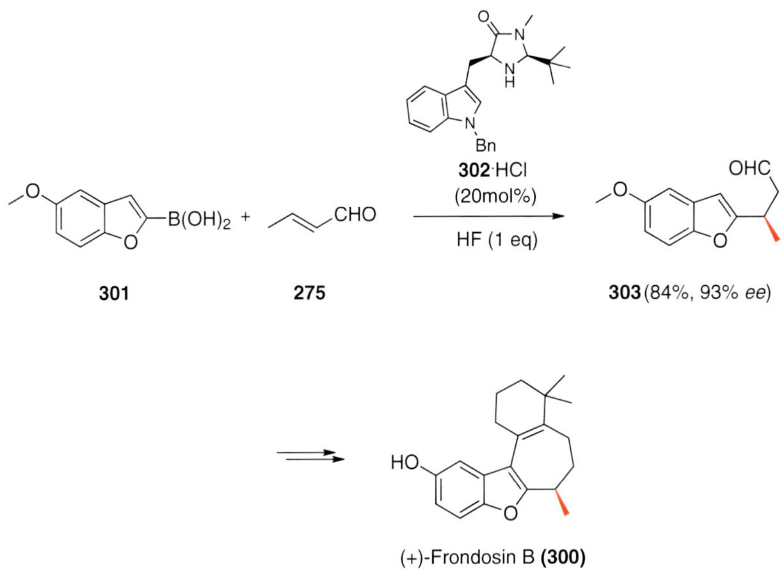

Scheme 68 *MacMillan's* three-step strategy for the synthesis of (+)-frondosin B (**300**)

3.2.4 Aza-Michael *Reactions*

Substituted nitrogen-containing heterocycles bearing a stereogenic center are another class of molecules that can be accessed using organocatalysis (*286*, *287*). One of the most fruitful approaches for this purpose is the intramolecular aza-*Michael* reaction (IMAMR) (*288*, *289*). *Fustero et al.* applied this strategy as the key step in the stereoselective synthesis of a variety of alkaloids (*288*) like the hemlock alkaloids (+)-coniine (**105**), (+)-sedamine (**304**), and (+)-allosedamine (**305**) (*290*). To obtain these piperidine alkaloids, the carbamate-tethered α,β-unsaturated aldehydes **306** could be converted efficiently into the chiral piperidines **307** *via* an intramolecular iminium-catalyzed aza-*Michael* addition using the secondary amine catalyst **276**. These aldehydes could then be used readily to access the alkaloids **105**, **304**, and **305** (Scheme 69).

The *Fustero* group recently also applied this strategy for the enantioselective total synthesis of (+)-angustureine (**308**) (*289*). (Scheme 70) Angustureine (**308**) is a tetrahydroquinoline alkaloid isolated from the bark of the Venezuelan shrubby tree *Galipea officinalis* (*291*). Carrying out the iminium-catalyzed intramolecular

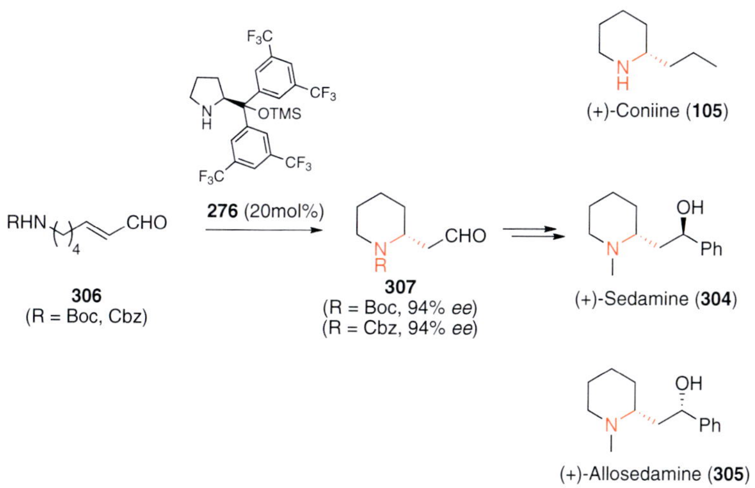

Scheme 69 Intramolecular aza-*Michael* reaction (IMAMR) for the syntheses of the hemlock alkaloids (+)-coniine (**105**), (+)-sedamine (**304**), and (+)-allosedamine (**305**) (*288*)

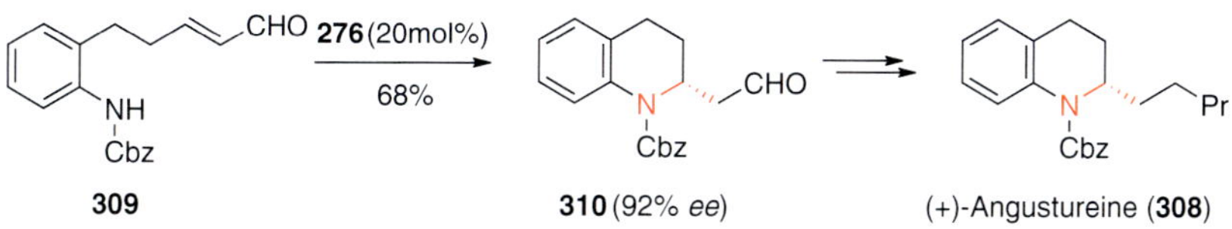

Scheme 70 IMAMR in the synthesis of (+)-angustureine (**308**)

aza-*Michael* addition on substrate **309**, the almost enantiopure tetrahydroquinoline **310** could be obtained in reasonable yield (68%). A series of standard functional group manipulations then gave access to the natural product **308** in a straightforward way (*289*).

3.2.5 Oxygenations

In 2005 *Jørgensen et al.* introduced a novel strategy for the asymmetric organocatalytic epoxidation of α,β-unsaturated aldehydes (*292*) (Scheme 71). Using secondary amine catalysts, α,β-unsaturated aldehydes could be epoxidized stereoselectively using H_2O_2 as the oxidant. Mechanistically, the reaction can best be

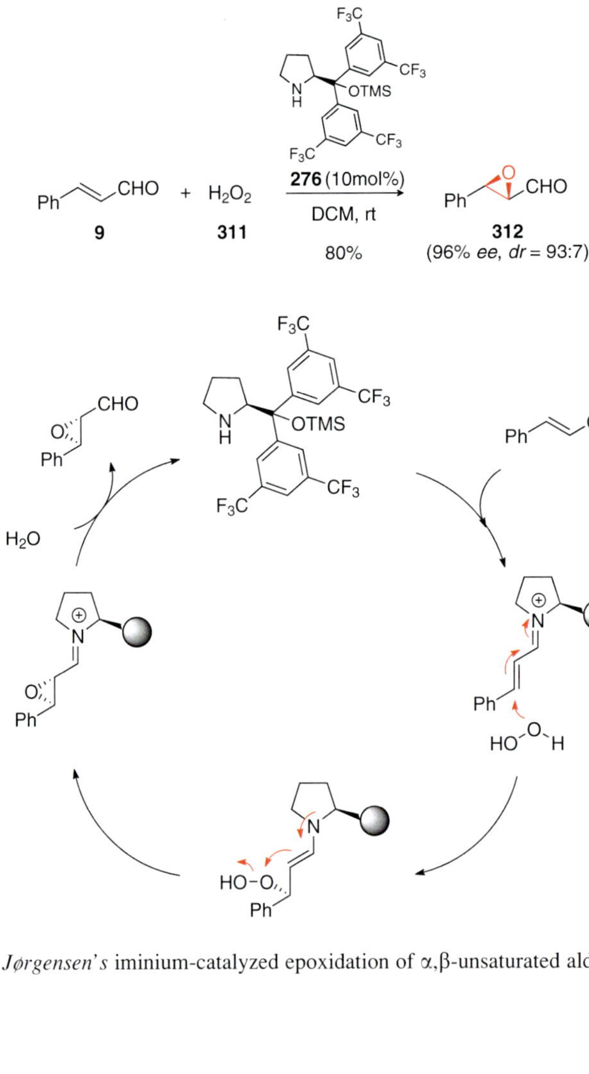

Scheme 71 *Jørgensen's* iminium-catalyzed epoxidation of α,β-unsaturated aldehydes (*292*)

described by an initial iminium-activation of the enal, followed by a conjugate addition of the peroxide and nucleophilic attack of the enamine intermediate at the electrophilic peroxygen atom to give an α,β-epoxidized iminium intermediate. Hydrolysis of the latter then liberates the target product (*292*). The procedure was found to give high yields and diastereoselectivities as well as very high enantioselectivities for a broad range of α,β-unsaturated aldehydes (*e.g.* aromatic, terpenoid) (*292*).

The *Nicolaou* group recently employed this method in their total synthesis of the fungal metabolite hirsutellone B (**313**) (*293*). Isolated from the insect pathogenic fungus *Hirsutella nivea* BCC 2594, the hirsutellones show impressive activities against *Mycobacterium tuberculosis*, the causative pathogen of tuberculosis (*294*, *295*). The organocatalytic epoxidation was carried out early in the construction of the tricyclic core of **313**. Using the (+)-citronellal-based iodo enal **314** as the substrate, epoxidation was carried out with amine **138** as the catalyst. The resulting α,β-epoxy aldehyde was converted directly further into the ester **315**. Intermediate **315** was then successfully transformed into the tricyclic fragment **316**, paving the way to yet another impressive natural product total synthesis from the *Nicolaou* group (Scheme 72) (*293*).

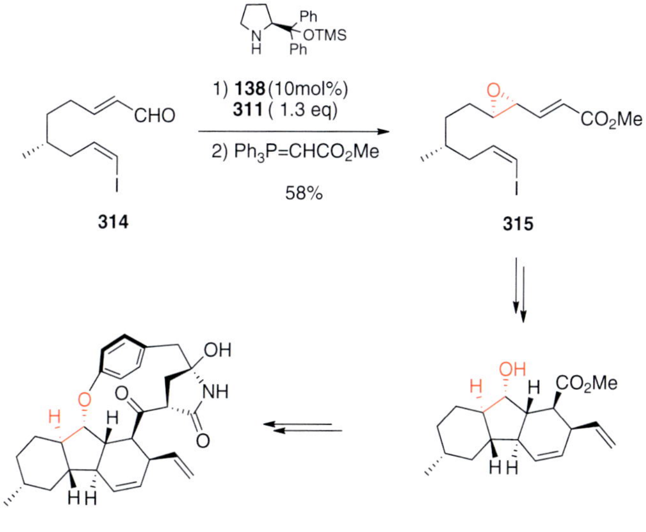

Scheme 72 Iminium-catalyzed epoxidation early in *Nicolaou's* total synthesis of hirsutellone B (**313**)

In 2007, the *Jørgensen* group reported a highly enantioselective procedure for the conjugate addition of oximes to α,β-unsaturated aldehydes in the presence of a chiral amine catalyst (*296*). This protocol was later expanded to the one-pot synthesis of optically active β-diols by carrying out a vinylogous addition of oxime **317** to different α,β-unsaturated aldehydes in the presence of catalyst **276** followed by a direct reduction of the primary addition product with $LiAlH_4$ (*297*).

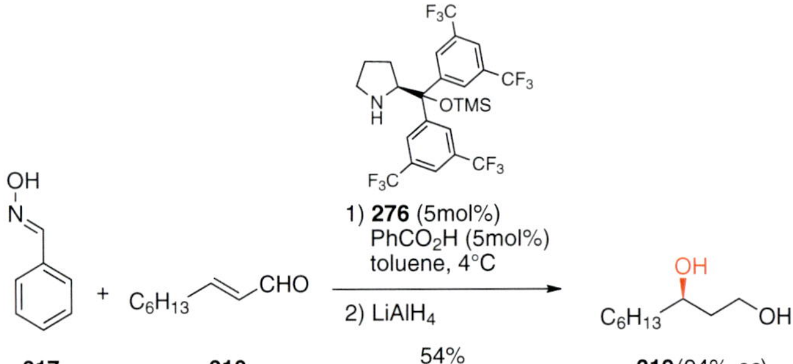

317 318 319 (94% *ee*)

Scheme 73 Iminium-catalyzed one-pot β-hydroxylation-reduction procedure for the syntheses of chiral β-diols

Employing aldehyde **318** as the electrophile, nonane-1,3-diol (**319**) could be obtained in reasonable yield (54%) and high enantioselectivity (94% *ee*) (Scheme 73). Compound **319** is the main constituent of an endogenous sex pheromone from the melon fly *Bactrocera cucurbitae* (*298*).

3.3 Iminium Catalyzed Organocascade Reactions

Cascade reactions have attracted considerable interest over the last years as they can facilitate the short-step preparation of complex structural motifs in good yields. It was already shown in Sect. 2.6 that enamine catalysis can be successfully employed in organocascade reactions to access complex natural products.

Recently *MacMillan et al.* have developed an iminium catalyzed organocascade approach to access a common intermediate in the synthesis of six different *Strychnos*, *Aspidosperma*, and *Kopsia* alkaloids (*299*). From a synthesis-oriented chemist's point of view, this work is rather elegant as it readily allows the easy and straightforward synthesis of a well-chosen key intermediate for the generation of structurally diverse natural products from simple starting materials. In this manner, the rapid synthesis of larger quantities of different alkaloids can be undertaken. This strategy was of course inspired by the fact that several *Strychnos*, *Aspidosperma*, and *Kopsia* alkaloids share a common biosynthetic intermediate that arises biosynthetically through an enzymatic cascade reaction starting from tryptamine and secologanin (*300*).

Employing the high potential of iminium ion activation, *MacMillan* and coworkers were able to directly access the key intermediate **320** in a single operation cascade reaction between the tryptamine-derived indole derivate **321** and propynal (**226**) in the presence of their trademark catalyst **227**. As illustrated in Scheme 74, the reaction is assumed to proceed *via* two iminium catalysis cycles (*Diels-Alder* cyclization first, followed by conjugate addition). It is worth noting that the authors considered either the possibility of intermediate **322** to directly enter the second

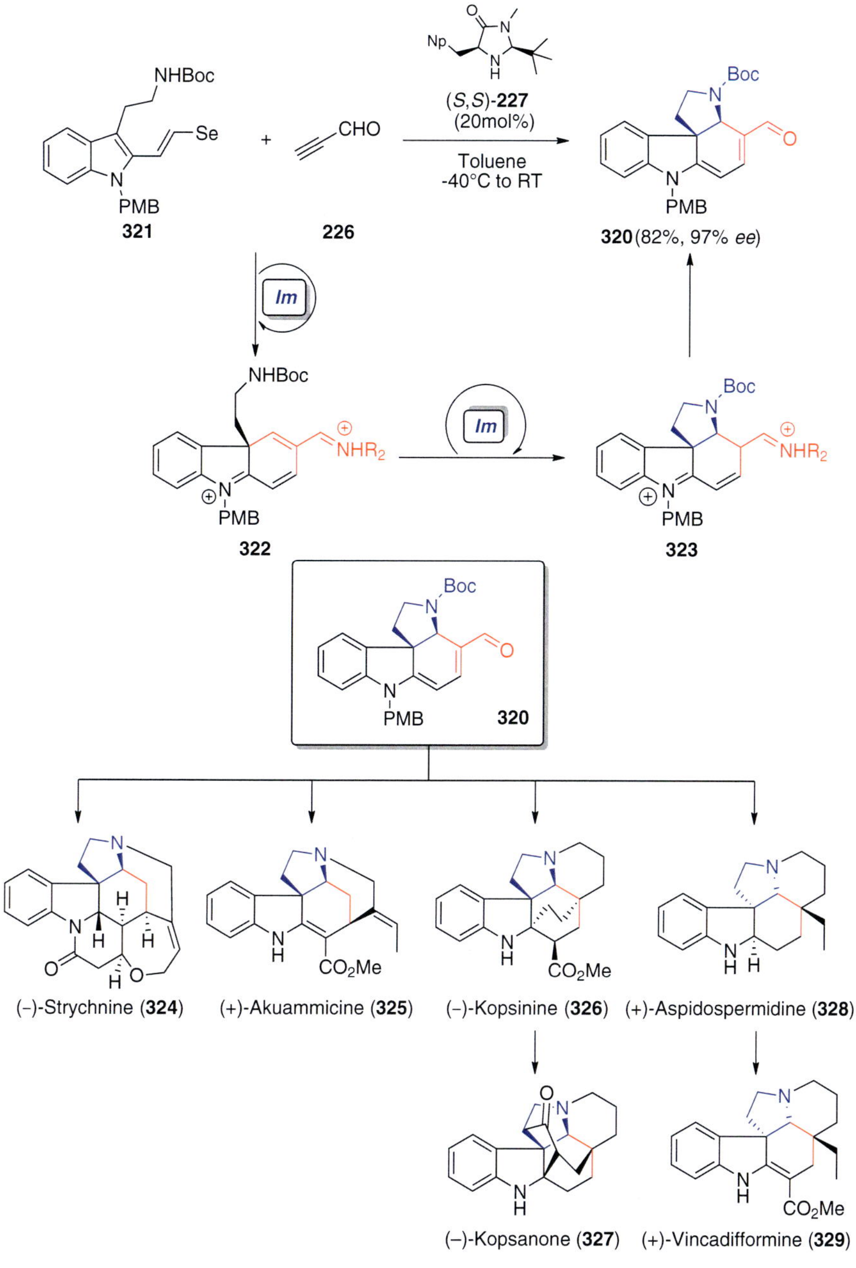

Scheme 74 Iminium-catalyzed synthesis of a key intermediate to access different *Strychnos*, *Aspidosperma*, and *Kopsia* alkaloids

cycle to give **323** or to undergo an alternative intramolecular cyclization – a *Brønsted* acid-catalyzed rearrangement that also gives **323** (not depicted in Scheme 74) (*299*).

The bioinspired intermediate **320** then served as the starting material to access six structurally different alkaloids in high yields: (−)-strychnine (**324**) (6.4% overall yield), (+)-akuammicine (**325**) (10%), (−)-kopsinine (**326**) (14%), (−)-kopsanone (**327**) (10%), (+)-aspidospermidine (**328**) (24%), and (+)-vincadifformine (**329**) (8.9%). The last two of these compounds were obtained by using the other enantiomer of catalyst **227** in the organocascade step. Remarkably, these syntheses are by far the shortest and the highest yielding ones published so far for these important biologically active compounds and the interested reader is referred to the original report by *MacMillan et al.* (*299*) for further details.

3.4 Synopsis

When summing up the recent achievements in iminium-activated natural product synthesis, the importance and versatility of this methodology cannot be overemphasized. Besides enamine catalysis, it is due particularly to the considerable achievements made in iminium catalysis that asymmetric organocatalysis has received so much attention over the last few years. As depicted in this chapter, the LUMO-lowering concept originally introduced by *MacMillan* has found widespread applications in natural product synthesis. In addition, it has been shown, that this activation mode works very well in cascade approaches. This methodology should become more widely utilized in the future (Table 2).

Table 2 Iminium catalysts employed in complex natural product syntheses

Catalyst	Product	References
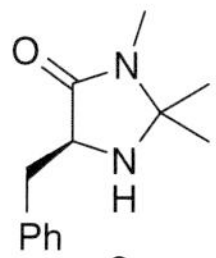 **8**	(+)-Hapalindole Q (**220**)	(*191*)
	(−)-Rhazinal (**288**)	(*261*)
	(−)-Rhazinilam (**289**)	(*261*)
	(−)-Leuconolam (**290**)	(*261*)
	(+)-*epi*-Leuconolam (**291**)	(*261*)
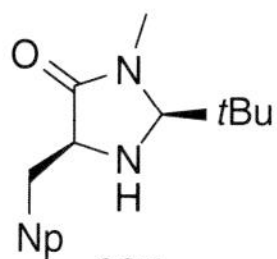 **227**	(+)-Minfiensine (**224**)	(*194*)
	(−)-Strychnine (**324**)	(*299*)
	(+)-Akuammicine (**325**)	(*299*)
	(−)-Kopsinine (**326**)	(*299*)
	(−)-Kopsanone (**327**)	(*299*)
	(+)-Aspidospermidine (**328**)	(*299*)
	(+)-Vincadifformine (**329**)	(*299*)
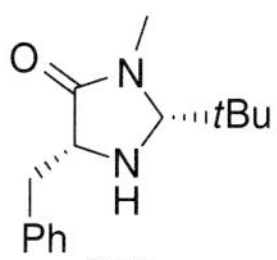 **228**	(+)-Minfiensine (**224**)	(*194*)
	Solanapyrone D (**235**)	(*195*)
	Amaminol A (**236**), B (**237**)	(*196*, *197*)
	Spiculisporic acid (**262**)	(*244*)
	(+)-Compactin (**269**)	(*245*)
	(−)-Flustramine B (**283**)	(*257*)
	(+)-Curcuphenol (**294**)	(*267*)
	Diazonamide A (**297**)	(*271*)
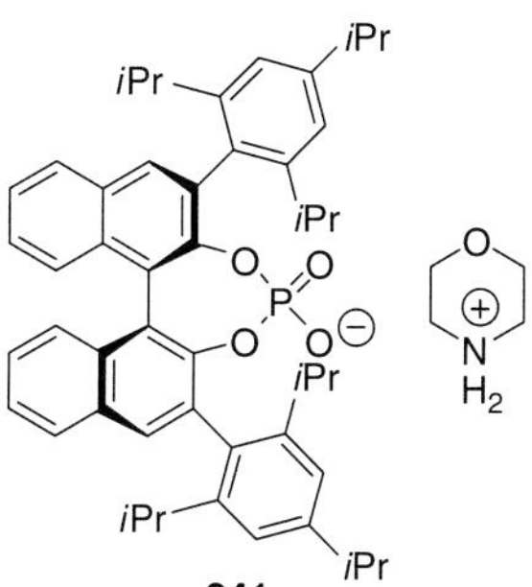 **241**	Pheromone (*R*)-**244**	(*210*)
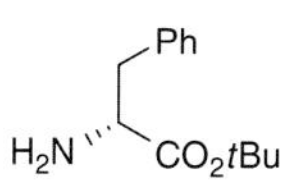 **246**	(−)-Platensimycin (**245**)	(*213*)
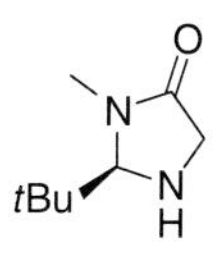 **253**	Dysideaproline E (**256**)	(*221*)

(continued)

Table 2 (continued)

Catalyst	Product	References
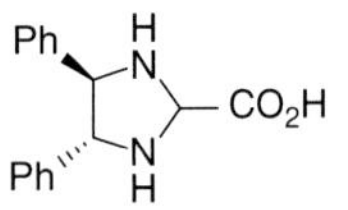 **261**	Warfarin (**257**)	(*230*)
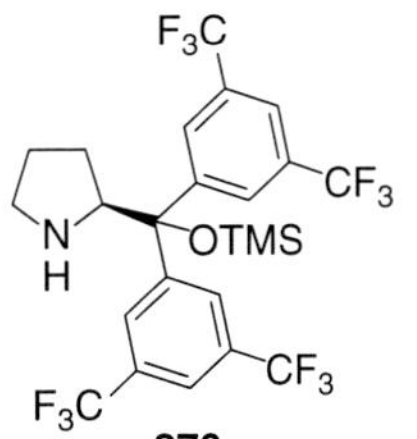 **276**	(+)-Fawcettimine (**273**) (+)-Coniine (**105**) (+)-Sedamine (**304**) (+)-Allosedamine (**305**) (+)-Angustureine (**308**) Pheromone **319**	(*132*) (*288*) (*288*) (*288*) (*289*) (*297*)
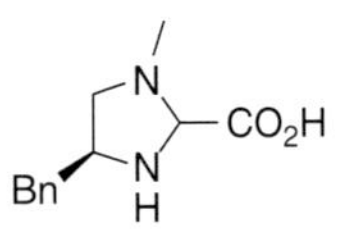 **282**	7-Deoxy-*trans*-dihydronarciclasine (**279**)	(*252*)
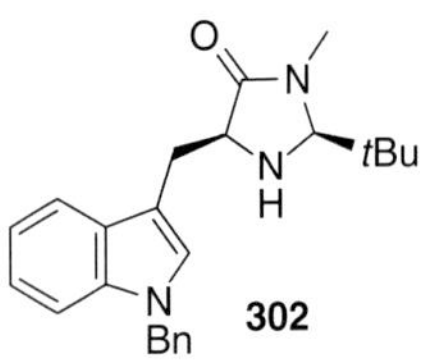 **302**	Frondosin B (**300**)	(*279*)
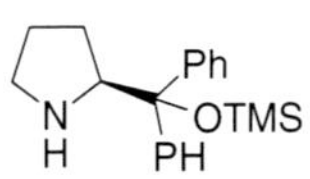 **138**	Hirsutellone B (**313**)	(*293*)

4 Combined Iminium-Enamine Catalyzed Approaches

It has been discussed in Sect. 2.6 that the combination of different organocatalytic activation modes for the efficient and short syntheses of complex structural moieties has attracted considerable interest over the last few years (*30*, *32*, *176–178*). As depicted in Schemes 48 and 49, the use of different enamine-catalysis based stereoselective transformations has given access to such highly functionalized and complex natural products like brasoside (**198**), littoralisone (**199**) (*115*), and callipeltoside C (**206**) (*184*), in an elegant and highly efficient way.

The main focus in this chapter will be on combined approaches using enamine catalysis and iminium catalysis especially in one-pot cascade reactions. As discussed in the following examples, the combined use of these two activation modes has led to the development of some of the most impressive and efficient organocatalytic natural product syntheses conducted so far (*301–305*).

4.1 Cascade Reactions Using a Single Organocatalyst

One of the first examples of combined stereoselective enamine-iminium catalysis in the synthesis of a natural product was reported by *Hong et al.* in 2006 (*301*). Using (*S*)-proline (**12**) as a catalyst (50%), a formal [3 + 3] cycloaddition of α,β-unsaturated aldehydes like crotonaldehyde (**275**) could be achieved. Mechanistically, this dimerization of **275** can be considered to proceed *via* a vinylogous *Michael* addition of an enamine-activated molecule of **275** to a second iminium-activated molecule of **275**, followed by an intramolecular aldol addition to give the diastereomeric cyclohexene carboxaldehydes **330** and **331** (Scheme 75). Of note, the initial *Michael*-addition proceeds with very high face selectivity whereas the aldol addition is less selective, thus giving two diastereomers (**330** and **331**) in good *ee*. The carboxaldehydes **330** and **331** were then converted successfully into (−)-isopulegol hydrate (**332**) and (−)-cubebaol (**333**). These 8-hydroxy-menthols have been reported recently to have biological activities like plant growth inhibitory and mosquito-repellent effects (*306*).

M. Waser, *Asymmetric Organocatalysis in Natural Product Syntheses*,
Progress in the Chemistry of Organic Natural Products 96,
DOI 10.1007/978-3-7091-1163-5_4, © Springer-Verlag Wien 2012

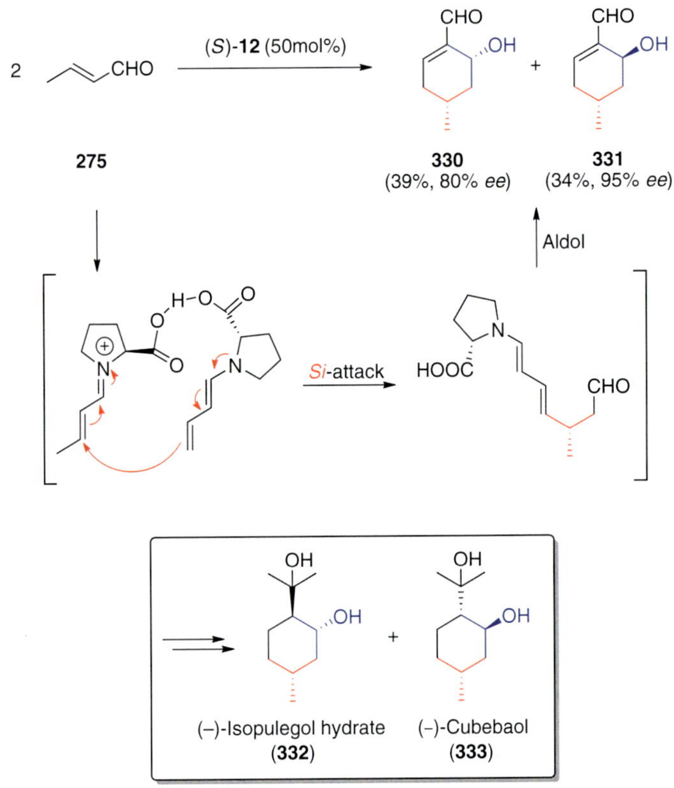

Scheme 75 Enamine-iminium cascade in the syntheses of (−)-isopulegol hydrate (**332**) and (−)-cubebaol (**333**)

A similar trimerization strategy of acrolein (**285**) was used recently by *Hong et al.* (Scheme 76) early in the total synthesis of montiporyne F (**334**) (*302*), a natural product isolated from the coral *Montipora* sp. exhibiting cytotoxic activity against human tumor cell lines (*307*). Importantly, trimerization of **285** in the presence of (*S*)-**12** and several additives always provided the racemic dialdehyde **335**. Although this is an example of a procedure where no stereodifferentiation

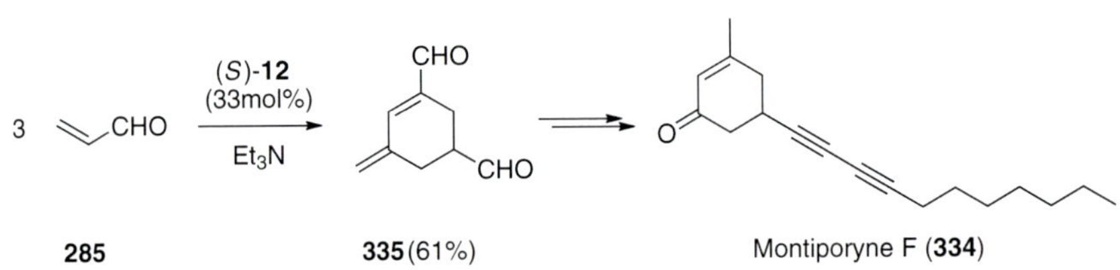

Scheme 76 Enamine-iminium cascade in the synthesis of montiporyne F (**334**)

could be achieved, it is noteworthy as it affords access towards cyclohexene carbdialdehydes that are not accessible so readily using conventional methods. Compound **335** could then be used to access the naturally occurring **334** in a straightforward manner. The absolute configuration of the natural product **334** has not yet been determined (*307*).

The *List* group reported recently an elegant total synthesis of the molluscicidal natural product (+)-ricciocarpin A (**336**) (*303*). The furanosesquiterpene lactone **336** was isolated from the liverwort *Ricciocarpos natans* and exhibits potent molluscicidal activity against the water snail *Biomphalaria glabrata*, a vector of schistosomiasis (*308*, *309*). The key steps in this synthesis approach included an organocatalytic one-pot conjugate reduction-*Michael* addition cascade followed by a $Sm(OiPr)_3$-catalyzed isomerization-*Tishchenko* cyclization (*303*). Starting from the enal **337**, a cascade reduction-*Michael* addition was achieved by treating **337** with 20 mol% of the *MacMillan* catalyst **228** and 1.1 equiv. of the *Hantzsch* ester **338**, furnishing the ketoaldehydes *cis*-**339** and *trans*-**339** in a 2:1 ratio and in excellent enantioselectivities (99% *ee* and 97% *ee*). Formation of **339** can be explained by an initial conjugate iminium-catalyzed reduction of the enal moiety and a direct intermolecular *Michael* addition of the resulting enamine intermediate **340** to the enone functionality. Subsequent treatment of this reaction mixture with $Sm(OiPr)_3$ not only causes a *cis-trans* isomerization, but also results in a highly diastereoselective *Tishchenko* cyclization reaction giving (+)-ricciocarpin A (**336**) as the *trans*-diastereomer exclusively (48% yield, 99% *ee*) (Scheme 77).

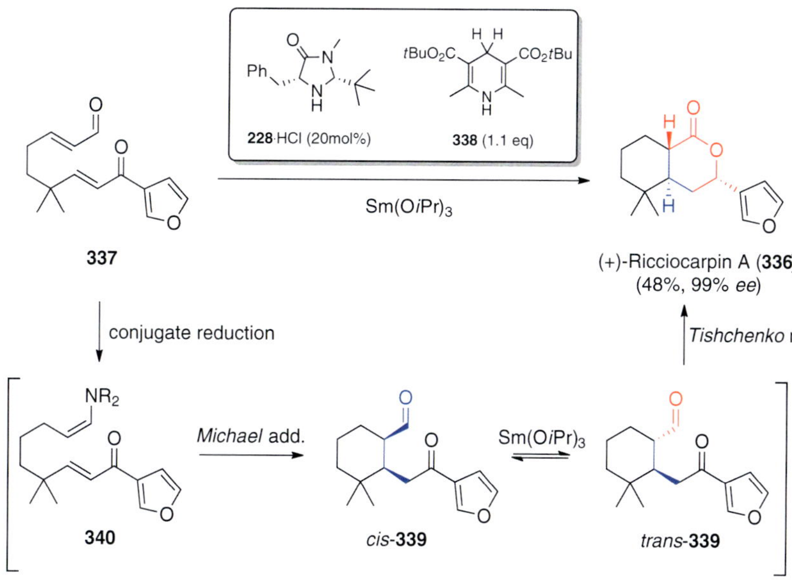

Scheme 77 Cascade reaction for the total synthesis of (+)-ricciocarpin A (**336**)

A three-component organocascade reaction was the key step in a recent synthesis of (+)-conicol (**341**) by *Hong et al.* (*305*). The benzo[*c*]chromene based (+)-**341** was isolated from the ascidian *Aplidium conicum* (*310*). Whereas no details about the bioactivity of conicol (**341**) have been reported so far, its isomer (+)-epiconicol (**342**) exhibits antiproliferative activity against human acute lymphoblastic leukemia cells and antibacterial activity against the *Gram*-positive bacterium *Micrococcus luteus* (*311*). The synthesis of the targeted hexahydro-6*H*-benzo[*c*]chromene skeleton of **341** could be achieved successfully by an organocatalytic three-component cascade reaction of the α,β-unsaturated nitro compound **343** and the enals **275** and **344** in the presence of the catalyst (*S*)-**138** (20 mol%) (Scheme 78). Mechanistically, the reaction is thought to proceed *via* an iminium-activated oxa-*Michael* addition of **343**–**344** initially, followed by an immediate enamine-catalyzed *Michael* addition to give **345**. Further *Michael* addition to crotonaldehyde **275** followed by a final intramolecular aldol condensation then results in compound **346**. Synthesis of the key intermediate **346** could either be achieved in a two-step protocol, isolating **345** or in a one-pot procedure giving **346** directly in 66% overall yield. This compound could then be employed successfully in the final steps to obtain (+)-conicol (**341**) (Scheme 78) (*305*).

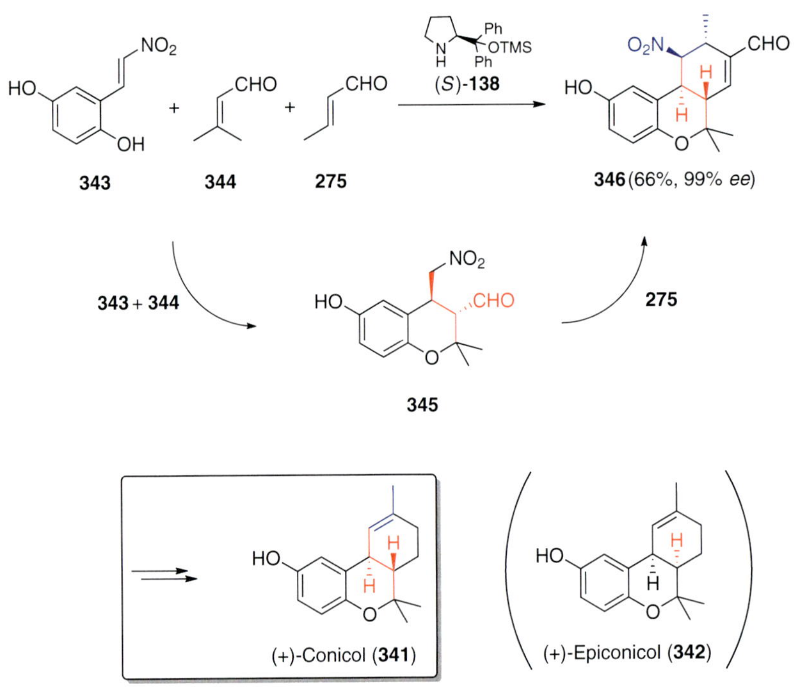

Scheme 78 Three-component cascade reaction in the total synthesis of (+)-conicol (**341**)

4.2 Organocascade Catalysis Using a Combination of Different Catalysts

The merging of two discrete organocatalytic cycles in a cascade fashion represents a very useful methodology for the syntheses of complex molecules in a short and efficient way (*304*, *312*). The concept of organocascade catalysis combines two or even more modes of substrate activation and allows reaction intermediates to shuttle between the different catalytic cycles. Thus, in contrast to a cascade reaction using just one catalyst, different catalysts can be combined to achieve a targeted transformation. Recently, *MacMillan et al.* (*304*) (Scheme 79) combined (*S*)-proline (**12**) and imidazolidinone (*S*,*S*)-**228** in the same reaction as a dual-catalyst

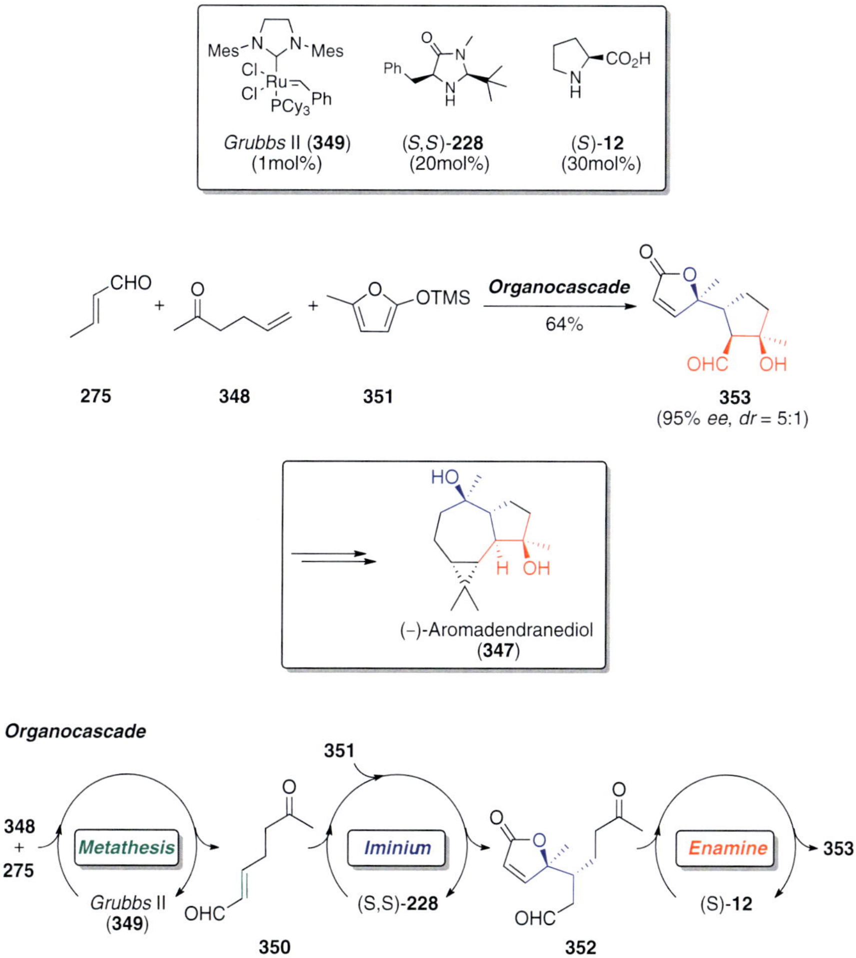

Scheme 79 Organocascade catalysis in the total synthesis of (−)-aromadendranediol (**347**)

system to carry out both iminium and enamine activation in a cascade fashion. Thus, using (*S*,*S*)-**228** to achieve exclusive iminium activation and (*S*)-(**12**) for enamine activation, the core of the naturally occurring sesquiterpene (−)-aromadendranediol (**347**) could be accessed in an elegant cascade reaction with excellent stereocontrol. (−)-Aromadendranediol (**347**) is a widely distributed sesquiterpene isolated both from the marine coral *Sinularia mayi* (*313*) and the leaves of *Xylopia brasiliensis* (*314*). The biological activity of **347** has not yet been widely studied, but it is known that this natural product is a constituent of plants found in Brazilian (*314*) and Chinese (*315*) folk medicines, which are used as sedatives, analgesics, and to treat lung inflammation. The first total synthesis of this structurally challenging target commenced with a cross-metathesis reaction between crotonaldehyde (**275**) and 5-hexene-2-one (**348**) (using *Grubbs*' second generation catalyst, **349**) to obtain the enal **350**. Subsequent addition of silyloxyfuran **351** and imidazolidinone (*S*,*S*)-**228** gave intermediate **352** and a final addition of (*S*)-proline (**12**) gave access to the target **353**. As depicted in Scheme 79, the whole sequence can be carried out in a one-pot triple-cascade sequence by adding the correct catalysts and reagents stepwise to give **353** in 64% yield (95% *ee*, *dr* = 5:1) (*304*). Further transformations then gave the natural product **347** in a rather short and high yielding sequence.

4.3 Synopsis

It has been shown in this chapter that combined iminium-enamine catalyzed approaches offer the potential to achieve complex (multistep) transformations in a single operational step, thus giving access to high structural complexity combined with excellent effectivity. In addition, it was pointed out also that organocatalysis can be used successfully when combined with metal catalysis (*e.g.* metathesis). Although these complex approaches have been pursued only for the last few years, the results obtained so far are very promising and thus this methodology should be applied more widely in the future (Table 3).

Table 3 Catalysts employed in combined iminium-enamine catalyzed natural product syntheses

Catalyst	Product	References
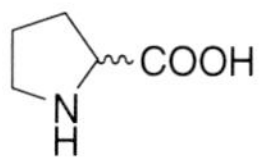 **12**	(−)-Isopulegol hydrate (**332**) (−)-Cubebaol (**333**) Montiporyne F (**334**) (−)-Aromadendranediol (**347**)	(*301*) (*301*) (*302*) (*304*)
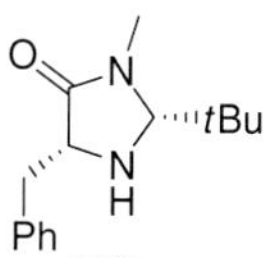 **228**	(+)-Ricciocarpin A (**336**) (−)-Aromadendranediol (**347**)	(*303*) (*304*)
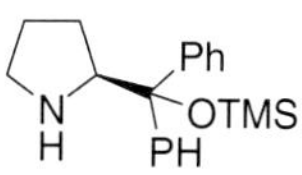 **138**	(+)-Conicol (**341**)	(*305*)

5 Singly Occupied Molecular Orbital (SOMO) Catalysis

Comparing the activation mode of iminium and enamine catalysis, iminium catalysis is based on a LUMO-activation mode of the electrophile whereas enamine catalysis is based on a HOMO-activation of the nucleophile. Keeping in mind the fact that enamine and iminium species are rapidly interconverted *via* a two-electron redox process (proton abstraction of an iminium species results in an enamine), *MacMillan* and co-workers reasoned that it should be possible to interrupt this equilibrium chemically by carrying out just a one-electron oxidation of an enamine. This would then generate a three-π-electron radical cation with a singly occupied molecular orbital (SOMO) that should be activated towards catalytic transformations (racemic or asymmetric) not possible using classical enamine or iminium activation (Scheme 80) (*316*).

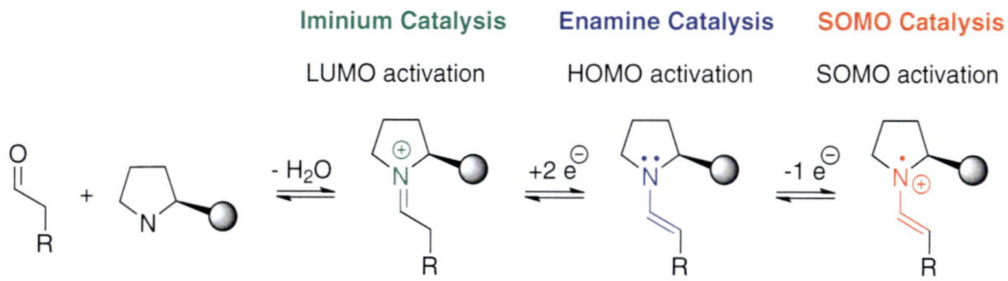

Scheme 80 Single-electron oxidation of a transiently formed enamine for the formation of a three-π-electron radical cation (SOMO activation)

In their seminal studies, *MacMillan et al.,* found that SOMO-activation of different aldehydes (*e.g.* **354**) in the presence of the chiral secondary amine catalyst **228** is possible by using 2 equiv. of ceric ammonium nitrate (CAN, **355**) as the stoichiometric oxidant. The SOMO-activated intermediate formed in this manner (**356**) reacted with allyltrimethylsilane (**357**) as a nucleophile to give the corresponding α-allylated aldehydes (like **358**) in high yields and enantioselectivities (Scheme 81). Accordingly, this methodology unites the unique reactivity of radicals with the high stereoselectivities that can be achieved using chiral secondary amine-based organocatalysts.

M. Waser, *Asymmetric Organocatalysis in Natural Product Syntheses*, Progress in the Chemistry of Organic Natural Products 96, DOI 10.1007/978-3-7091-1163-5_5, © Springer-Verlag Wien 2012

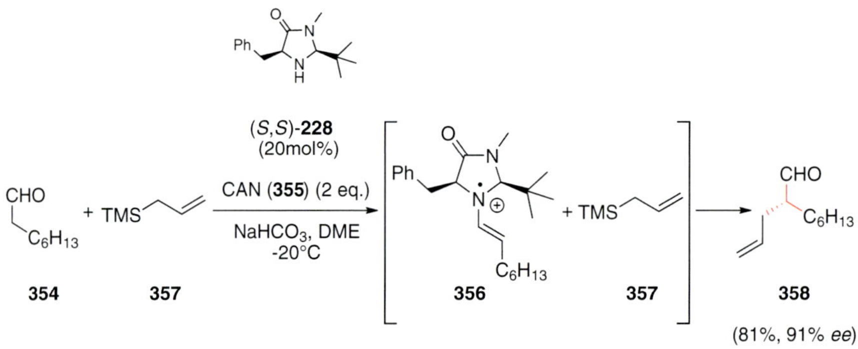

Scheme 81 SOMO-activation for the stereoselective α-allylation of aldehydes

Following this milestone report, the *MacMillan* group developed this efficient *Umpolung*-methodology further and successfully applied it to the stereoselective addition of different π-electron neutral or π-electron rich nucleophiles to the electrophilic α-position of the intermediate aldehyde-derived radical species, giving access to products that are otherwise obtained by conventional methods only with difficulty (*317–320*).

Although it has been shown in several studies over the last few years that this type of methodology possesses great potential for the synthesis of a variety of different structural motifs, the application to natural product synthesis has so far been rather limited (*321*, *322*). This is somewhat surprising, but it may be expected that it will take some time until these new reaction methodologies really find their place in the synthesis-oriented chemist's standard toolbox for natural product synthesis.

5.1 *Friedel-Crafts* Reactions

The first to apply SOMO-activation in a complex total synthesis was the *Nicolaou* group in 2009 (*321*). Using an intramolecular *Friedel-Crafts*-type α-arylation of an aldehyde tethered to an electron-rich aromatic group, an elegant and efficient total synthesis of demethyl calamenene (**359**) could be achieved. Demethyl calamenene is a naturally occurring tetralin-based cytotoxic agent active against the A549 human adenocarcinoma cell line (*323*). Structurally, one key feature of this compound is a benzylic stereogenic center, which can be built with high enantioselectivity by carrying out an intramolecular α-arylation of aldehyde **360** in the presence of catalyst (*R*,*R*)-**228** and 2 equiv. of CAN (**355**). Accordingly, this example underscores the benefits of an *Umpolung*-strategy in making the α-position of a carbonyl group a highly suitable electrophile for stereoselective *Friedel-Crafts*-type reactions. As depicted in Scheme 82, the reaction is considered to proceed *via* initial formation of the enamine **361**, followed by a single electron transfer (SET)

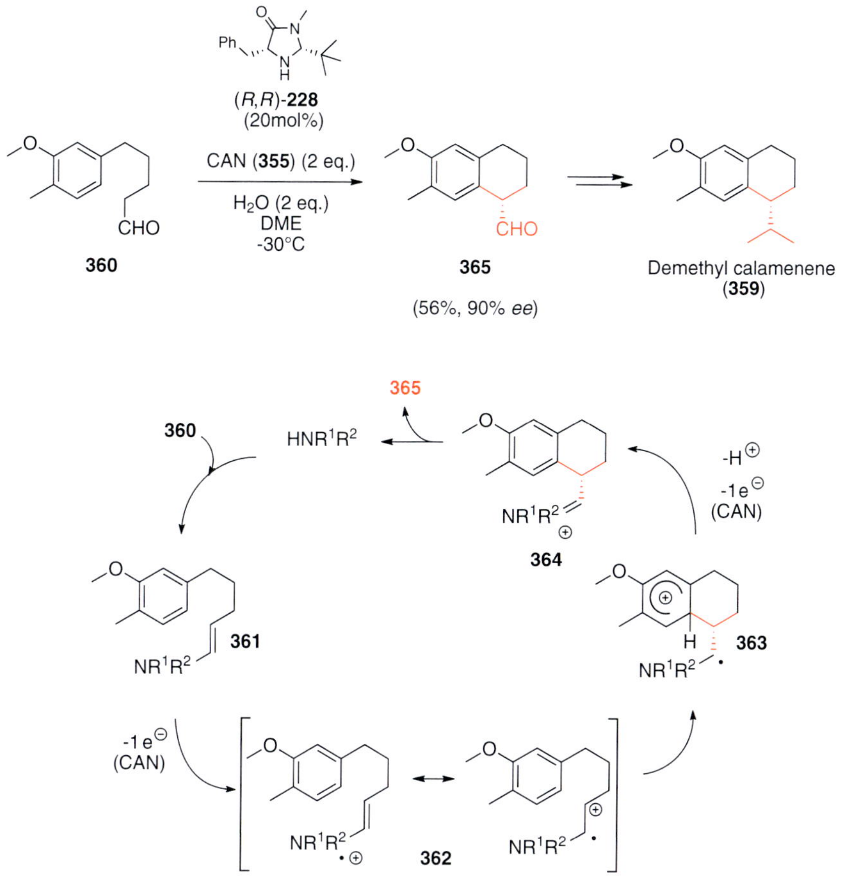

Scheme 82 SOMO-activated *Friedel-Crafts*-type α-arylation in the synthesis of demethyl calamenene (**359**)

oxidation to give the mesomeric radical intermediates **362**, which immediately form the σ-complex **363** with high face selectivity. Proton abstraction and further SET-oxidation gives the iminium **364**, which, upon hydrolysis, results in tetralin **365**. Additional functional group transformations can then give access to demethyl calamenene (**359**) in just three more steps (*321*).

5.2 Epoxide Formation

MacMillan et al. developed a SOMO-activation protocol for the stereoselective α-chlorination of aldehydes and for the formation of terminal epoxides starting from aldehydes (*317*). This method was employed recently by the *Christmann*

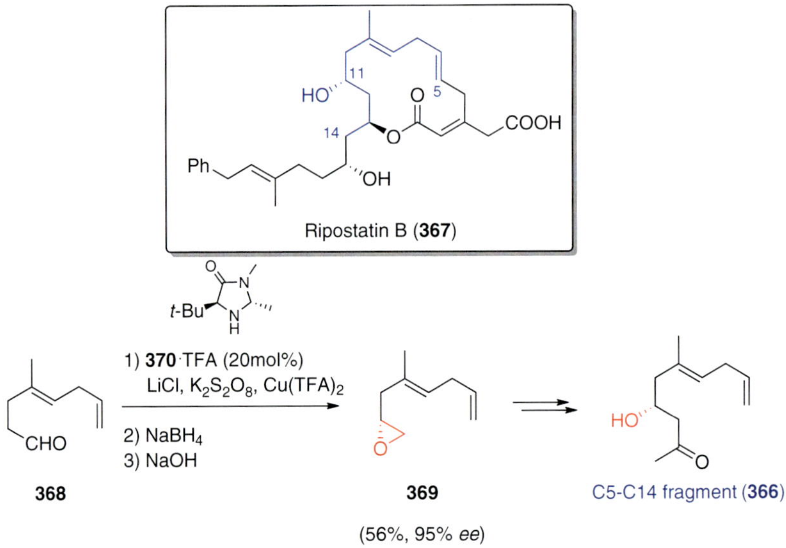

Scheme 83 SOMO-activated epoxide formation in the synthesis of the C-5–C-14 fragment (**366**) of ripostatin B (**367**)

group to synthesize the C-5–C-14 fragment **366** of the potent RNA polymerase inhibitor ripostatin B (**367**) (*322*, *324*). Although **366** was not transformed further into the polyketide **367**, this report is notable as it demonstrates the utilization of easily available terpene-based starting materials for the construction of complex polyketides. The organocatalytic introduction of the C-11 stereogenic center was achieved by applying *MacMillan's* epoxidation protocol on the epoxygeranyl acetate-derived aldehyde **368**. The targeted terminal epoxide **369** could thereby be obtained in a satisfactory yield and a high enantioselectivity by carrying out an α-chlorination with LiCl, using the imidazolidinone catalyst **370** in the presence of an oxidant combination of copper(II) trifluoroacetate (Cu(TFA)$_2$) and $K_2S_2O_8$, followed by an immediate reduction with NaBH$_4$ and intramolecular S_N-substitution. The ripostatin B fragment **366** was then obtained in two additional steps, illustrating the usefulness of this simple terpene feedstock-based protocol to access complex polyketide architectures (Scheme 83).

5.3 Synopsis

SOMO catalysis is one of the most recent developments in asymmetric organocatalysis and its considerable promise has been shown thus far in a limited number of complex natural product syntheses. As this unprecedented *Umpolung*-methodology enables stereoselective transformations that are otherwise barely possible, further applications in demanding syntheses can be expected soon by others (Table 4).

Table 4 Organocatalysts employed in SOMO-activated natural product syntheses

Catalyst	Product	References
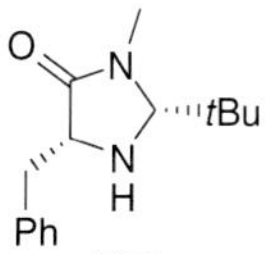 **228**	Demethyl calamenene (**359**)	(*321*)
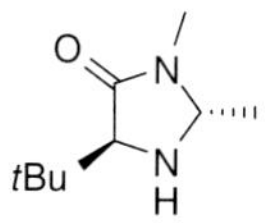 **370**	Ripostatin B fragment **366**	(*322*)

6 Asymmetric Phase-Transfer Catalysis

The term "phase-transfer catalysis" was introduced in 1971 by *Starks*, explaining the critical role of tetraalkylammonium or phosphonium salts to promote reactions between two substances located in different immiscible phases (*325*). Over the years, the use of achiral quaternary ammonium salts as phase-transfer catalysts (PTCs) has attracted widespread interest not only in academia but also for industrial applications (*326*, *327*). Some of the most important benefits of phase-transfer catalysis are simple experimental conditions, which are usually easily scalable, in addition to mild reaction conditions, and the use of inexpensive and environmentally friendly reagents and solvents.

While extensive investigations concerning the use of achiral onium species were carried out quite soon after the first reports in the 1970s, the development of asymmetric phase-transfer catalysis has progressed significantly more slowly (*328–332*). From a historical point of view, this is rather surprising as the first examples by *Wynberg* date back to the 1970s (*333*) and the first highly enantioselective application using a chiral *Cinchona* alkaloid-derived catalyst was reported back in 1984 by a group at Merck Research Laboratories (*19*) (Scheme 84).

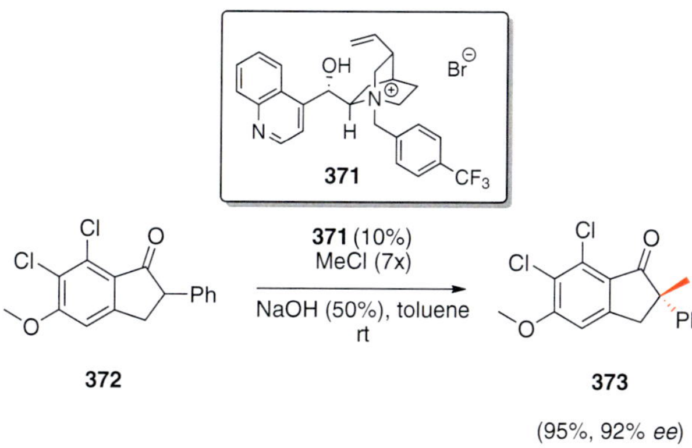

Scheme 84 Phase-transfer catalyzed asymmetric α-alkylation developed by Merck scientists

M. Waser, *Asymmetric Organocatalysis in Natural Product Syntheses*,
Progress in the Chemistry of Organic Natural Products 96,
DOI 10.1007/978-3-7091-1163-5_6, © Springer-Verlag Wien 2012

By using the *Cinchona* alkaloid-derived quaternary ammonium bromide **371**, a stereoselective methylation of the phenylindanone **372** was achieved under biphasic conditions, thus representing one of the first examples of such a highly stereoselective organocatalytic asymmetric transformation. These types of transformations may be conducted in a stereoselective fashion using more commonly employed methods only with great difficulty.

Triggered by these pioneering studies, several different structurally well-defined quaternary ammonium salts (Q^+X^-) have been developed and investigated thoroughly and employed successfully for the catalysis of different asymmetric reactions, mainly involving anionic species (*328*, *329*). As shown in the next few sections, some of these reactions have also been used successfully for the synthesis of complex natural products.

6.1 Asymmetric α-Alkylations

One class of application that readily highlights the enormous potential of asymmetric phase-transfer catalysis is the stereoselective α-alkylation of different carbanion nucleophiles, in particular enolates. Although these types of transformations are most important in organic chemistry, there are still only a limited number of stereoselective catalytic methods available and the use of chiral PTCs represents one of the most versatile strategies to achieve such transformations. One example of special interest is the asymmetric α-alkylation of glycine *Schiff* base **374** (Scheme 85)

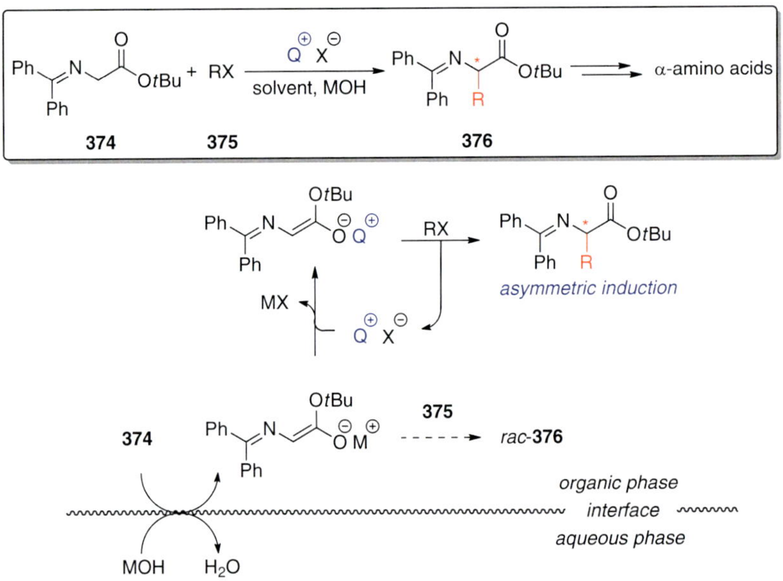

Scheme 85 Phase-transfer catalyzed syntheses of (non-) natural amino acids starting from the achiral glycine *Schiff* base **374**

with various electrophiles (**375**) giving access to natural and unnatural α-amino acids on hydrolysis of the alkylation products (**376**). *O'Donnell* and co-workers first reported this reaction in 1989 by using *Cinchona* alkaloid-derived catalysts (*334*), and, over the years, this has become a benchmark reaction for new chiral ammonium salt catalysts (*335–340*).

The accepted mechanistic explanation for stereo-differentiation suggests formation of a contact ion pair between the cationic chiral ammonium species and the enolate. As illustrated in Scheme 85, the key step therein is a cation exchange of the initially formed achiral enolate (which would lead to formation of racemic **376** upon alkylation) with the PTC to give a chiral ion pair, and with the sterically defined ammonium group shielding one face of the enolate while exposing the other one towards the electrophile. In terms of the exact geometry of the enolate, it is worth noting that both the (*E*)- and the (*Z*)-enolate have been postulated to be involved in these types of reactions (*335–337*).

Phase-transfer catalyzed asymmetric α-alkylations of achiral carbonyl precursors have been used to access complex natural products, as exemplified by the synthesis of (−)-esermethole (**377**) of *Wong et al.*, a compound which is an important precursor in the synthesis of the naturally occurring anticholinesterase agent (−)-physostigmine (**378**) (*341*). Accordingly, a phase-transfer catalyzed cyanomethylation of oxindole **379** with **380** in the presence of the quaternary ammonium salt **381** gave the key intermediate **382** in good yield and reasonable enantioselectivity. Compound **382** could then be cyclized, resulting in a simple and stereoselective synthesis of (−)-esermethole (**377**) (Scheme 86).

Another application was reported by *Andrus et al.* in the total synthesis of the hydroxy ketones kurasoin A (**383**) (*342*) and B (**384**) (*343*). These two compounds

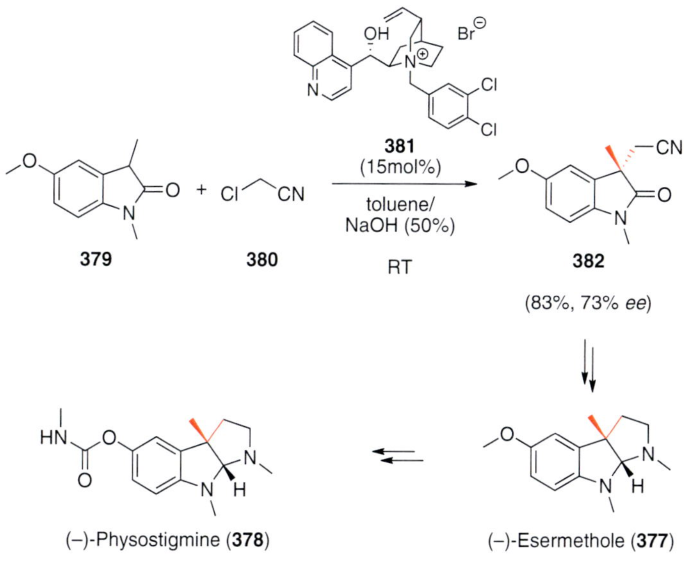

Scheme 86 Phase-transfer catalyzed oxindole alkylation

were isolated during a search for protein farnesyltransferase inhibitors from the fungus *Paecilomyces* sp. (*344*). They show potential as novel anticancer drug leads since the aromatic substituents can be modified easily in a systematic way around the central α-hydroxy ketone core. An asymmetric glycolate alkylation of the dimethoxyacetophenone **385** with the benzyl bromide **386** in the presence of the cinchonidinium catalyst **387** gave the O-protected α-hydroxy ketone **388** in high yield and good enantioselectivity and permitted access to kurasoin A (**383**) in a few additional steps (*342*). The synthesis of kurasoin B (**384**) could be accomplished starting from the benzyloxyacetyl imidazole **389** on reaction with the bromo compound **390** in the presence of the biscinchonidinium dimethylnaphthalene catalyst **391**. This alkylation furnished the targeted intermediate **392** in excellent yield and enantioselectivity, thus paving the way to obtain **384** in a high overall yield of 34% (nine steps) (Scheme 87).

Bengamides B (**393**) and Z (**394**) are naturally occurring caprolactams belonging to a larger family of 24 members, which have been isolated from coral species and

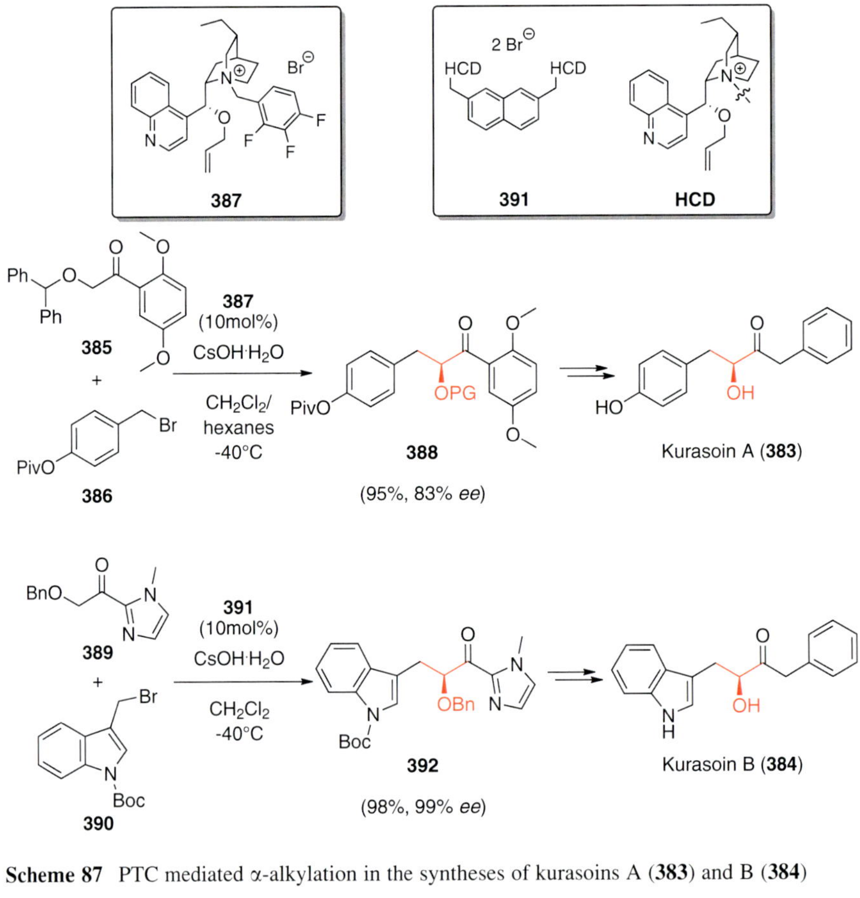

Scheme 87 PTC mediated α-alkylation in the syntheses of kurasoins A (**383**) and B (**384**)

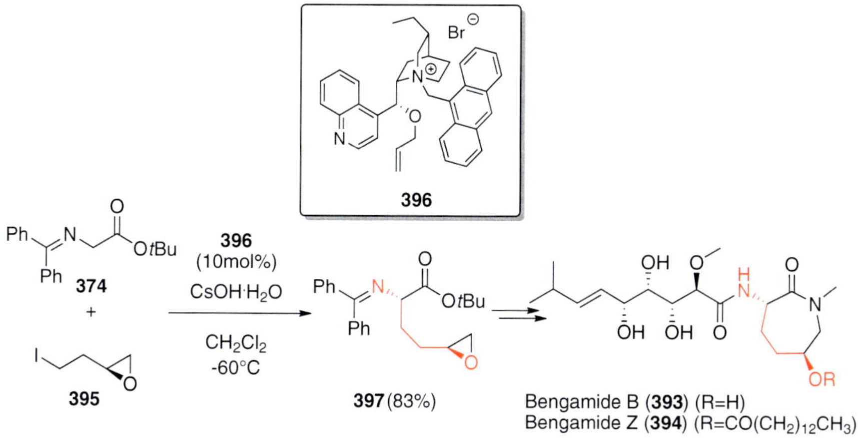

Scheme 88 Asymmetric α-alkylation in the syntheses of bengamides B (**393**) and Z (**394**)

marine sponges (*345*). Bengamide B (**393**) has attracted considerable interest as it showed a unique profile in the 60 cell-line panel of the U.S. National Cancer Institute when compared to standard antitumor agents (*346*). The required configuration of the caprolactam part was introduced *via* a phase-transfer catalyzed alkylation of the glycine *Schiff* base **374** with the chiral epoxide based electrophile **395** as recently demonstrated by *Boeckman et al.* (*346*). Using the *Cinchona* alkaloid-based catalyst **396**, compound **397** was obtained as a single diastereomer, serving as a useful intermediate to accomplish the total syntheses of both bengamide B (**393**) and bengamide Z (**394**) (Scheme 88).

A phase-transfer catalyzed α-alkylation of **374** was also one of the key steps in the first total synthesis of (−)-antofine (**398**) by *Kim et al.* (*347*). The naturally occurring phenanthroindolizidine alkaloid (−)-antofine (**398**) is a highly potent cancer cell growth inhibitor with IC_{50} values in the low nanomolar range (*348*). The stereogenic center could be installed by reacting **374** with electrophile **399** in the presence of the dimeric catalyst **391** (Scheme 89). With the key intermediate **400** in hand, final manipulations to obtain the natural product **398** were then achieved in a straightforward manner (*347*).

In 2003, the group of *Shibasaki* reported an elegant synthesis of aeruginosin 298-A (**401**), a potent serine protease inhibitor isolated from a blue-green alga (*349*), using chiral phase-transfer catalyzed alkylations in the syntheses of three fragments (*350*). Aeruginosin 298-A has a tetrapeptide-like structure including nonstandard amino acids, thus presenting an appropriate motif for phase-transfer catalyzed amino acid synthesis.

As depicted in Scheme 90, *Shibasaki's* two-center tartaric acid-derived catalysts **402** and **403** worked well for the installation of the stereogenic centers

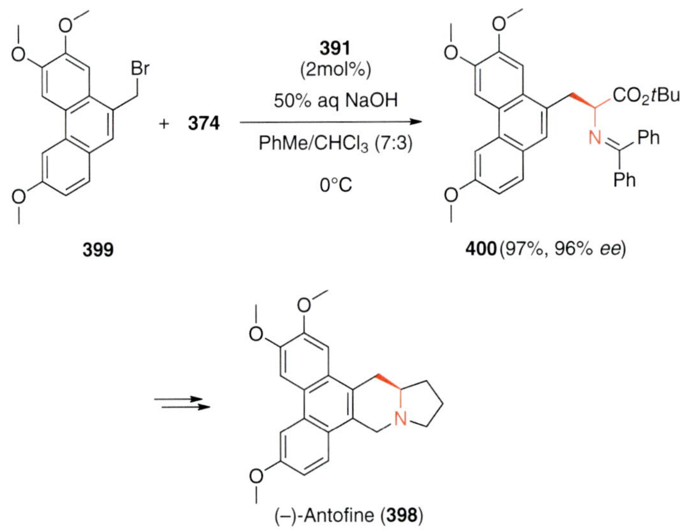

Scheme 89 First total synthesis of (−)-antofine (**398**)

of the amino acid fragments **405**, **407**, and **409** by appropriate alkylation of *Schiff* base **374** with the corresponding electrophiles **404**, **406**, and **408**, in the presence of the matching enantiomer of the catalyst. One interesting fact concerning this methodology is the importance of the counteranion X^- in these alkylation reactions. While catalyst **402** (X = I) gave fragment **409** in a reasonable 79% yield and 91% *ee*, the use of BF_4^- as the counteranion (catalyst **403**) gave an even better yield (85%) and enhanced the enantioselectivity slightly (93% *ee*). These amino acid-based fragments could then be assembled successfully to obtain aeruginosin 298-A (**401**) in a reasonably short timeframe and in a high yielding sequence (Scheme 90) (*350*).

The asymmetric α-alkylation of glycine *Schiff* base **374** was found also to be useful for the syntheses of a variety of aroylalanine derivatives like kynurenine (**410**). Kynurenine (**410**) is an important metabolite in the oxidative cleavage of L-tryptophan, a pathway appearing to play an important role in a variety of fundamental biological processes like cell growth/cell division, thus making it a promising target with respect to drug development (*351*). *Lygo et al.* (*352*) carried out a stereoselective α-alkylation of **374** with allyl bromide **411** to obtain the Boc-protected amino acid **413** after protecting group exchange. Compound **413** was then transformed into aroylalanine derivatives like kynurenine (**410**) in a few more steps, thus resulting in a reliable procedure to access this important naturally occurring tryptophan metabolite (Scheme 91).

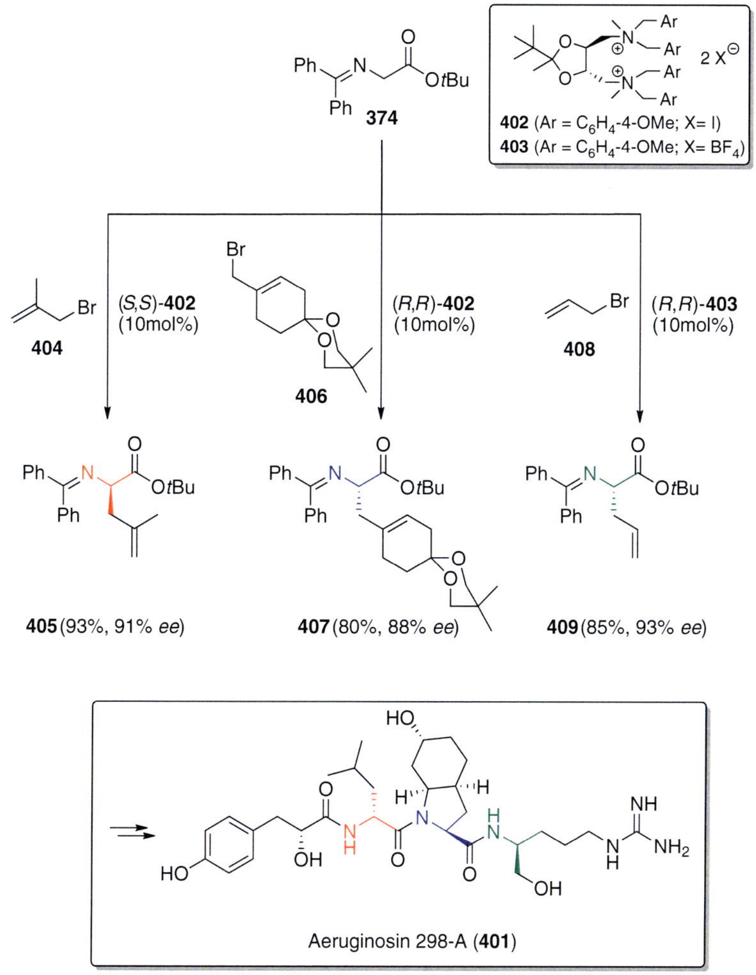

Scheme 90 *Shibasaki's* synthesis of aeruginosin 298-A (**401**)

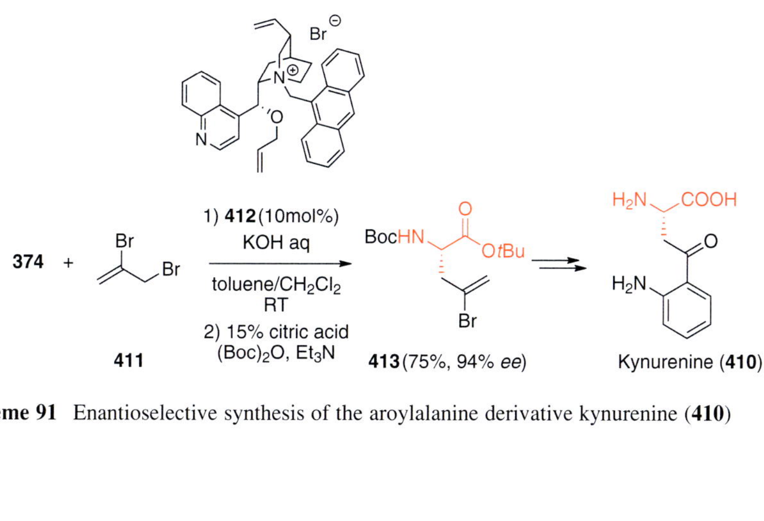

Scheme 91 Enantioselective synthesis of the aroylalanine derivative kynurenine (**410**)

6.2 *Phase-Transfer Catalyzed* Michael *Additions*

The paramount importance of *Michael* additions as versatile C–C bond forming transformations was discussed in some detail earlier in this volume. Thus, it is not surprising that, besides the use of chiral PTCs in asymmetric α-alkylation reactions, their use for stereoselective *Michael* additions is one of the most carefully investigated reactions in asymmetric phase-transfer catalysis (*328*, *329*). Accordingly, the additional use of this methodology in asymmetric total synthesis has been reported on several occasions.

(+)-Triptoquinone A (**414**) is a diterpenoid quinone isolated from the Chinese herb *Tripterigium wilfordii* var. *regelii*, possessing interleukin-1 inhibitory properties (*353*). An asymmetric synthesis of **414** using a phase-transfer catalyzed *Michael* addition key step was reported by *Shishido et al.* in 1994 (*354*). Using catalyst **415**, the required quaternary stereogenic center could be installed by reacting the tetralin derivative **416** with *Michael* acceptor **417**. The intermediate product **418** underwent a subsequent aldol condensation in the presence of catalytic amounts of [18]crown-6, thus giving the annulation product **419** in good yield and enantioselectivity. After this *Michael* addition/*Robinson* annelation step, the total synthesis of (+)-triptoquinone A (**414**) required standard functional group manipulations (Scheme 92), providing an elegant and very efficient method to access this complex natural product.

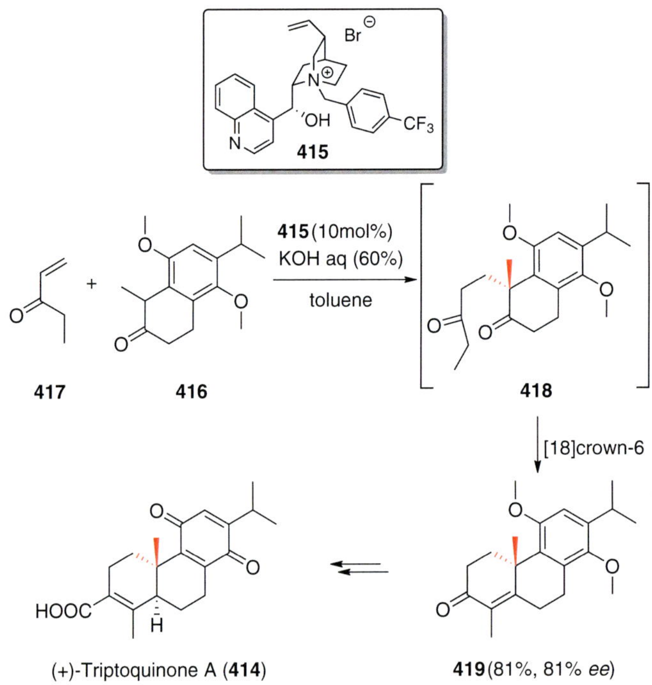

Scheme 92 Phase-transfer catalyzed *Michael* addition/*Robinson* annelation strategy in the synthesis of (+)-triptoquinone A (**414**)

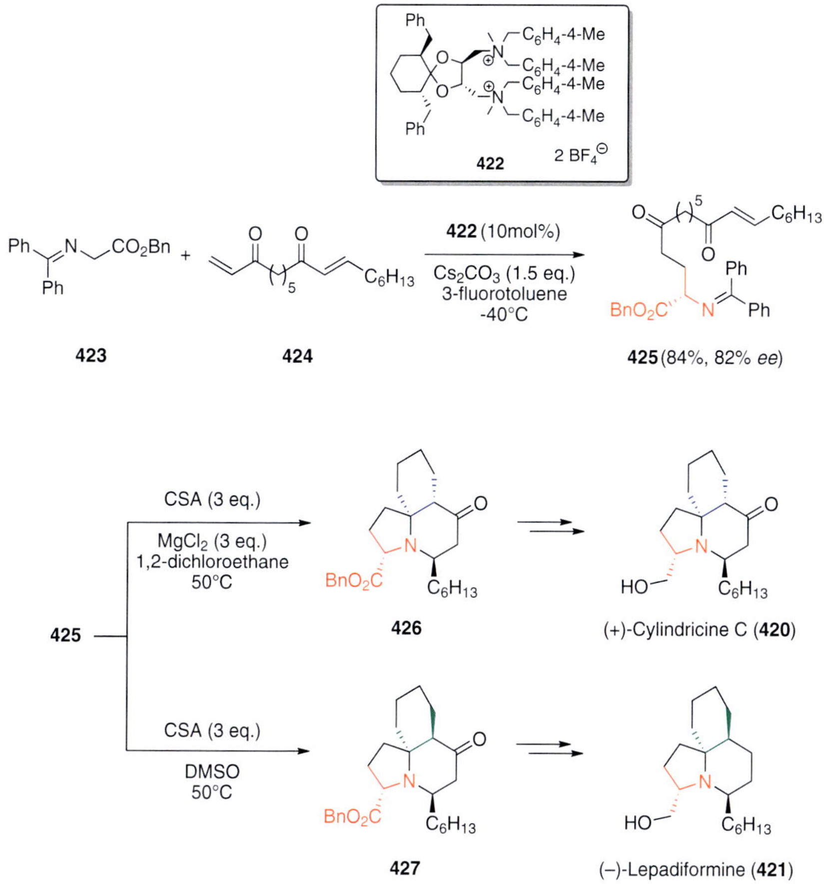

Scheme 93 Phase-transfer catalyzed *Michael* addition in the syntheses of the tricyclic alkaloids (+)-cylindricine C (**420**) and (−)-lepadiformine (**421**)

Shibasaki et al. demonstrated (Scheme 93) the applicability of their tartrate-based diammonium salt catalysts for the syntheses of the alkaloids (+)-cylindricine C (**420**) and (−)-lepadiformine (**421**) (*355–357*). The tricyclic cylindricines were isolated from the marine ascidian *Clavelina cylindrica* by *Blackman et al.* and these compounds exhibit bioactivity against a DNA-repair-deficient yeast strain as well as growth inhibition of murine leukemia and human solid-tumor cell lines (*358–360*). Lepadiformine (**421**) was isolated by *Biard et al.* from the marine tunicates *Clavelina lepadiformis* and from *Clavelina moluccensis* and has been shown to exhibit moderate cytotoxic activity against various tumor cell lines (*361*). By applying an ammonium salt **422**-catalyzed addition of *Schiff* base **423** to *Michael* acceptor **424**, the key intermediate **425** was obtained in good yield and with reasonable enantiomeric excess. Compound **425** could then be used to obtain selectively either the cylindricine C precursor **426** or the lepadiformine synthon **427** in a very efficient tandem cyclization reaction by choosing the optimum reagents. The impressively short total synthesis of (+)-cylindricine C (**420**) could be achieved in only two additional steps, whereas the synthesis of the tricyclic intermediate **427** represents a formal total synthesis of (−)-lepadiformine (**421**) (*355–357*).

It is worth noting that the *Shibasaki* group also developed a highly useful protocol for asymmetric *Mannich*-type reactions (*356*), which was used successfully in the synthesis of biologically active molecules like nemonapride, an antipsychotic agent developed by Yamanouchi Pharmaceutical. Since such a useful application of this methodology is not a *bona fide* natural product synthesis, a detailed discussion is beyond the scope of this contribution and the interested reader is referred to *Shibasaki's* original report (*356*).

6.3 Alkylative Dearomatization-Annulation Reaction

Although chiral ammonium salts have proven their utility in catalyzing a variety of different reactions on numerous occasions, applications in natural product synthesis have so far been limited mainly to α-alkylations and *Michael*-type reactions. Nevertheless, over the past several years increased efforts to broaden the scope of chiral PTCs towards even more complex reactions have been undertaken (*362–364*).

Recently, the group of *John A. Porco, Jr.* employed a *Cinchona* alkaloid-derived dimeric PTC to catalyze an alkylative dearomatization-annulation process to build up the highly functionalized adamantane core of hyperibone K (**428**) (*362*). Hyperibone K (**428**) was isolated from the perennial herb *Hypericum scabrum* and shows moderate activity as an inhibitor of human tumor cell replication (*365*). The *Porco* group found the dimeric catalyst **429** to be the best suited among a series of different *Cinchona*-based ammonium salts to promote the key-dearomatization/annulation reaction between the prenylated benzophenone clusiaphenone B (**430**) and enal **431**. This protocol gives the highly functionalized adamantane-based **432** in good yield (71%) and high enantioselectivity (90% *ee*) (Scheme 94). In addition, NMR studies showed that the starting material **430** occurs in a dearomatized enolate form mainly under standard reaction conditions, thus causing formation of a contact ion pair with the PTC followed by a cascade of highly face-selective C–C bond forming reactions.

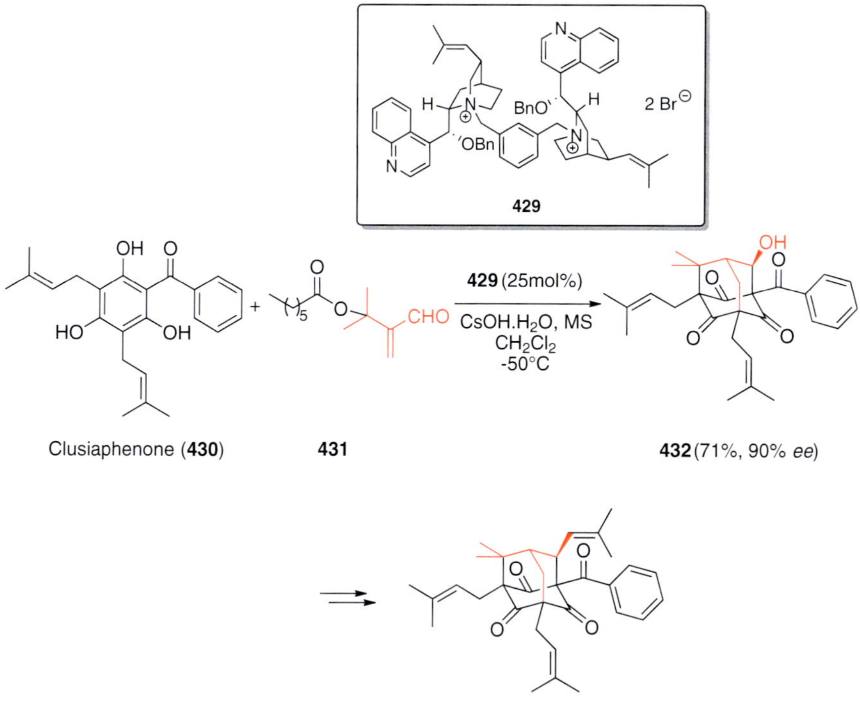

Clusiaphenone (**430**) **431** **432** (71%, 90% *ee*)

Hyperibone K (**428**)

Scheme 94 Phase-transfer catalyzed alkylative dearomatization-annulation strategy to access hyperibone K (**428**)

6.4 Synopsis

Asymmetric phase-transfer catalysis is a method that has for almost three decades proven its high utility. Although its typical application is for (non-natural) amino acid synthesis, over the years other types of applications have been reported. The unique capability of quaternary ammonium salts to form chiral ion pairs with anionic intermediates gives access to stereoselective transformations that are otherwise very difficult to conduct using metal catalysts or other organocatalysts. Thus, this catalytic principle has created its own very powerful niche within the field of asymmetric catalysis. As can be seen in Table 5 below, the privileged catalyst structures are mostly *Cinchona* alkaloid-based, whereas the highly potent *Maruoka*-type catalysts have so far not been applied routinely to complex natural product total synthesis.

Table 5 Asymmetric quaternary ammonium salts used in complex natural product syntheses

Catalyst	Product	References
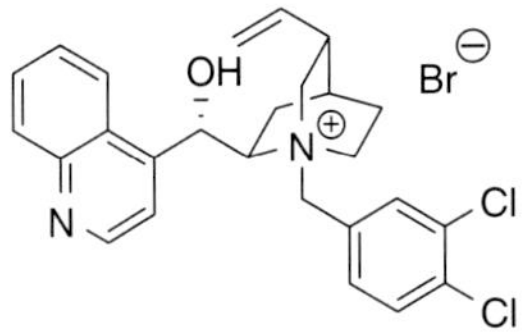 **381**	(−)-Esermethole (**377**)	(*341*)
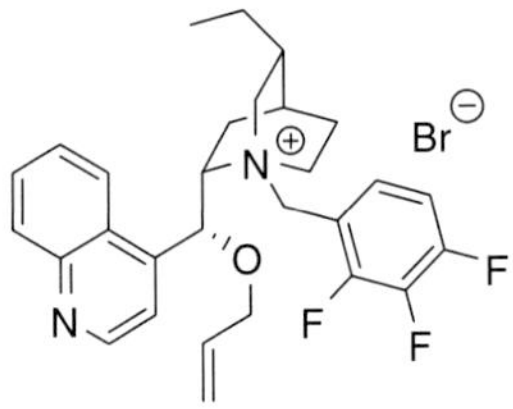 **387**	Kurasoin A (**383**)	(*342*)
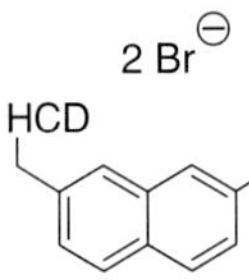 **391** 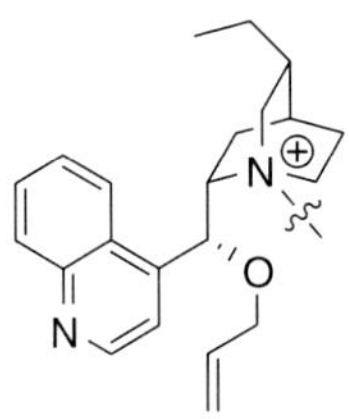 **HCD**	Kurasoin B (**384**) (−)-Antofine (**398**)	(*343*) (*347*)
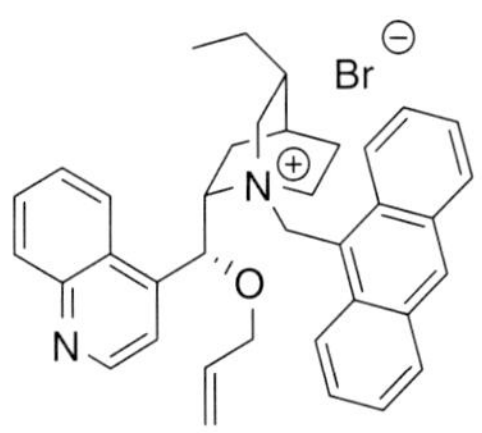 **396**	Bengamide B (**393**) Bengamide Z (**394**)	(*346*) (*346*)
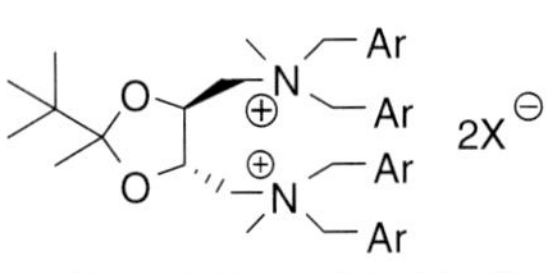 **402** (Ar = C_6H_4-4-OMe; X = I) **403** (Ar = C_6H_4-4-OMe; X = BF_4)	Aeruginosin 298-A (**401**)	(*350*)

(continued)

Table 5 (continued)

Catalyst	Product	References
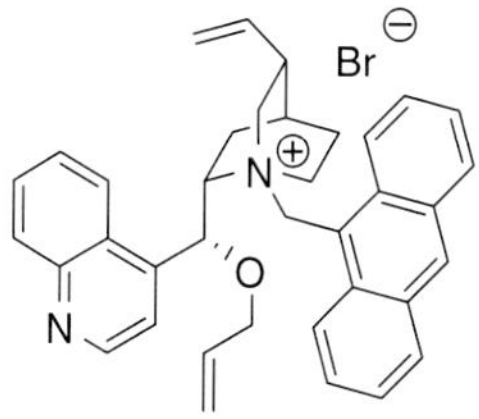 **412**	Kynurenine (**410**)	(*351*)
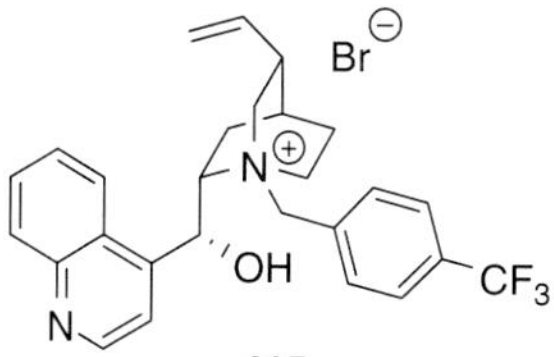 **415**	(+)-Triptoquinone A (**414**)	(*354*)
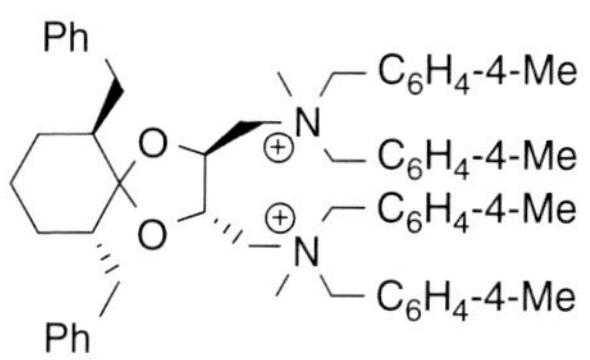 **422** 2 $BF_4^{\ominus}$	(+)-Cylindricine C (**420**) (−)-Lepadiformine (**421**)	(*355, 357*) (*355, 357*)
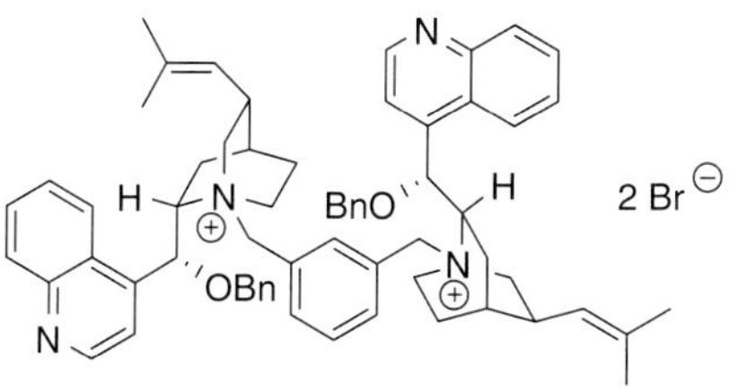 **429**	Hyperibone K (**428**)	(*362*)

7 Chiral *Brønsted* Acids and Hydrogen Bonding Donors

Hydrogen bonding is one of the most important attractive forces in Nature. In addition to its importance as a structural determinant, hydrogen bonding also plays a crucial role in catalysis. As hydrogen bonding to an electrophile decreases the electron density of this species, activation towards a nucleophilic attack is achieved. This principle is employed successfully by Nature's catalysts, the enzymes, for the acceleration of a wide range of chemical processes. Mimicking this strategy, organic chemists have begun only recently to appreciate the tremendous potential offered by hydrogen bonding as a catalytic principle for electrophile activation using small-molecule catalysts (*366–373*).

From a historical point of view, it is surprising that organic chemists did not start investigating this mode of catalysis until the 1990s (*366*), because the first report of *Wassermann* discussing the beneficial effect of protic additives such as carboxylic acids and phenols to accelerate the cycloaddition of cyclopentadiene with benzoquinone appeared as long ago as 1942 (*374*). However, members of the scientific community did not seem to realize the high potential of this methodology and instead the use of *Lewis* acids to accelerate organic reactions like *Diels-Alder* reactions (first reported in 1960 by *Yates* and *Eaton* (*375*)) has resulted in much more attention. It is not the purpose of this contribution to discuss the pros and cons of either method, but without doubt over the years both methodologies, *Lewis* acid and *Brønsted* acid catalysis, have proven their potential to catalyze a variety of different reactions in an achiral or chiral fashion, making them some of the most versatile catalytic procedures available to chemists.

Looking at the mode of activation, one should consider two commonly accepted mechanisms: (a) specific acid catalysis and (b) general acid catalysis. While specific acid catalysis refers to the reversible protonation of the electrophile with a strong acid in a pre-equilibrium step prior to nucleophilic attack, general acid catalysis involves the proton transfer or hydrogen bonding activation to the transition state in the rate-determining step (*e.g.* nucleophilic attack), usually under weakly acidic or neutral conditions (Scheme 95) (*366*).

Realizing the high potential of both strong *Brønsted* acids and weaker hydrogen bond donors as (chiral or achiral) small-molecule catalysts to activate electrophiles

M. Waser, *Asymmetric Organocatalysis in Natural Product Syntheses*,
Progress in the Chemistry of Organic Natural Products 96,
DOI 10.1007/978-3-7091-1163-5_7,

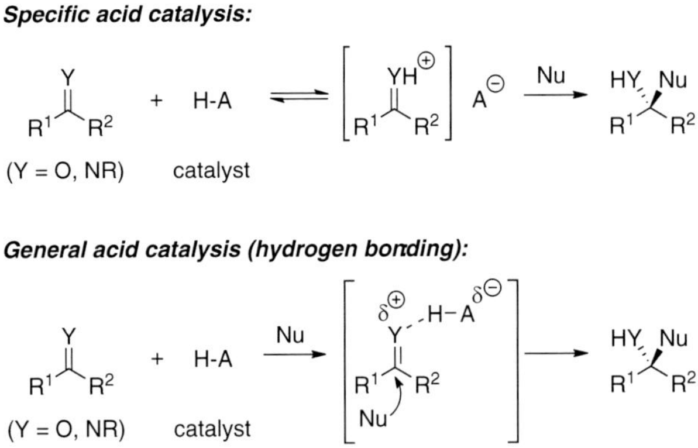

Scheme 95 Simplified mechanistic differences between specific and general acid catalysis

to facilitate different transformations, a variety of catalysts have been introduced successfully over the last few years (*366–372*). Like the other activation modes presented so far, asymmetric hydrogen bonding and *Brønsted* acid catalysts have proven their worth in several highly demanding natural product syntheses. In contrast to the previous chapters of this volume, the present treatment is divided according to the different types of catalysts employed and not according to the types of reactions carried out.

7.1 Chiral Phosphoric Acids

As protons are the simplest and most easily available *Lewis* acids available to catalyze organic transformations, the use of a chiral *Brønsted* acid combines the potential of proton catalysis with asymmetric induction achieved through the choice of the proper counteranion. It is without doubt that chiral phosphoric acids have been the most successfully used chiral *Brønsted* acids so far (*369*, *371*). With respect to the chiral backbone employed therein, binaphthyls have been found to be the most promising substances for this purpose.

In 2006, the group of *Rueping* reported a highly enantioselective organocatalytic transfer hydrogenation of substituted quinolines to access different plant alkaloids (*376*). Using only 2 mol% of phosphoric acid catalyst **433** in combination with the *Hantzsch* ester **247** as the hydride source, the conversion of the 2-substituted quinolines **434** into the corresponding tetrahydroquinolines **435** occurred in a single step procedure with excellent yields and enantioselectivities (Scheme 96). Subsequent reductive N-methylation then gave access to the tetrahydroquinoline alkaloids (+)-galipinine (**436**), (+)-cuspareine (**437**), and (−)-angustureine (**308**). Isolated from the Angostura tree *Galipea officinalis* (*291*, *377*), these alkaloids show inhibitory activity against *Mycobacterium tuberculosis* (*378*).

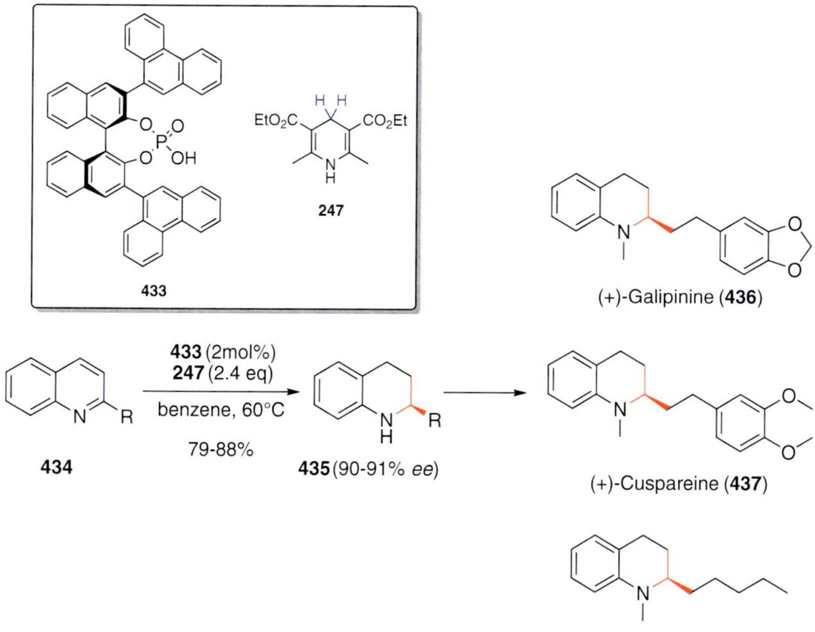

Scheme 96 Chiral phosphoric acid-catalyzed transfer hydrogenation in the syntheses of different biologically active tetrahydroquinoline alkaloids

The *Rueping* group was also able to expand this methodology to the enantioselective transfer hydrogenation of pyridine derivatives (*379*). Using the modified phosphoric acid catalyst **438** in combination with the hydride donor **247**, several trisubstituted pyridine derivatives were reduced successfully to the corresponding chiral piperidines in high selectivity and with good functional group tolerance (*379*). It is of note that the anthracene-based catalyst **438** was found to be more efficient in this case than the phenanthrene-based **433**. As a proof of concept of the applicability of this method for the synthesis of natural products, the skeleton of the pumiliotoxin family was obtained successfully by reduction of pyridine **439** under the conditions developed to furnish the corresponding hexahydroquinolinone **440**, a well-known key intermediate in the synthesis of *diepi*-pumiliotoxin C (**441**) (*380*) (Scheme 97). Pumiliotoxin C, first isolated from *Dendrobates pumilio*, is an interesting alkaloid obtained from poison dart frogs and a potent neurotoxin that acts as a noncompetitive blocker for acetylcholine receptor channels and therefore has attracted considerable attention from a pharmaceutical standpoint (*381*). This explains the attention given to the syntheses of structurally similar analogues or epimers.

Hiemstra et al. reported recently an asymmetric phosphoric acid-catalyzed *Pictet-Spengler* reaction of benzyltryptamines to gain access to chiral β-carbolines with good yields and selectivities (*382*). The versatility of this protocol was

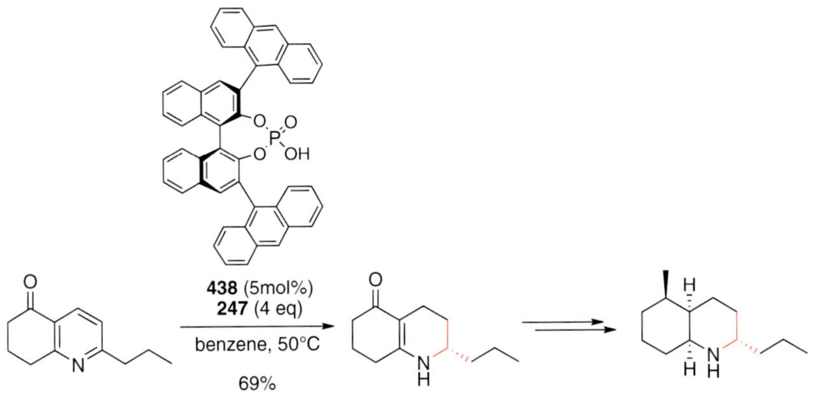

Scheme 97 Chiral phosphoric acid-catalyzed transfer hydrogenations of pyridines exemplified in the synthesis of *diepi*-pumiliotoxin C (**441**)

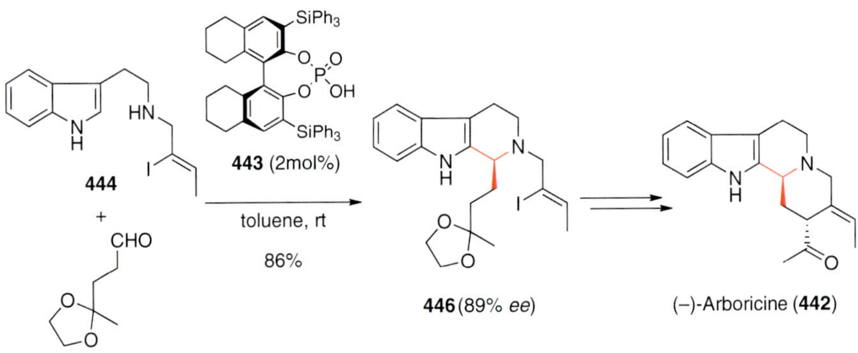

Scheme 98 Chiral phosphoric acid-catalyzed *Pictet-Spengler* reaction in the synthesis of (−)-arboricine (**442**)

demonstrated successfully in a short and scalable synthesis of (−)-arboricine (**442**) (*383*). Arboricine (**442**) is a deplancheine-type tetracyclic indole alkaloid isolated from the leaves of *Kopsia arborea*, a tree native to tropical Asia and NE Australia, by *Kam et al.* It shows moderate ability to reverse multi-drug resistance in vincristine-resistant KB (VJ300) cells (*384*). Using BINOL-based phosphoric acid **443** (2 mol%), the enantioselective *Pictet-Spengler* reaction between tryptamine derivative **444** and the protected oxopentanal **445** proceeded in high yield and with reasonably good enantioselectivity. Further functional group manipulations and an intramolecular cyclization furnished the natural product **442** in 33% overall yield (6 steps) (Scheme 98). Moreover, this procedure was

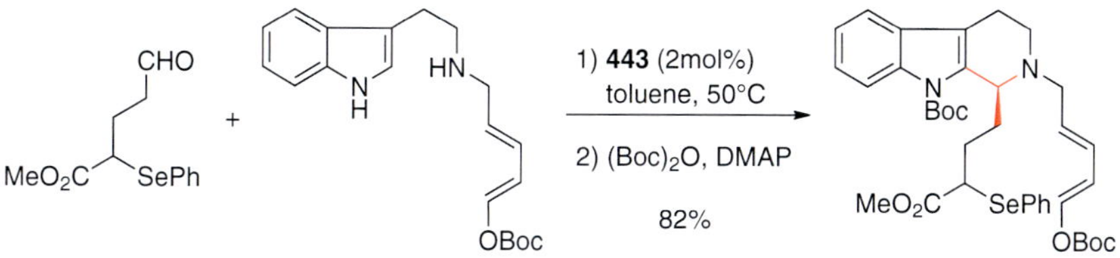

448 **449** **450** (84% *ee*)

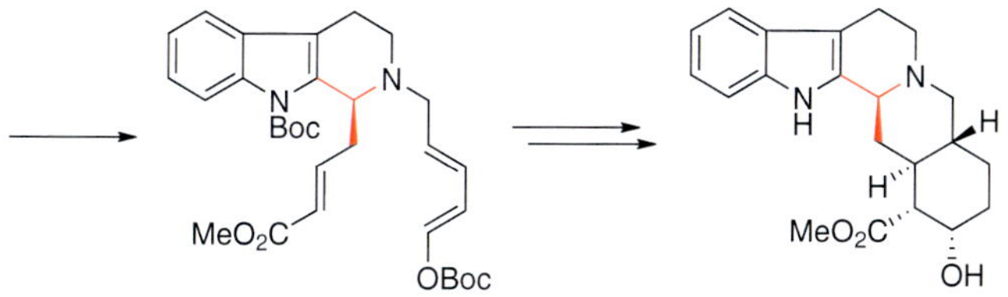

451 (+)-Yohimbine (**447**)

Scheme 99 Chiral phosphoric acid-catalyzed *Pictet-Spengler* reaction in the synthesis of (+)-yohimbine (**447**)

found to be readily useful up to a 10 mmol scale, thus giving access to reasonable amounts of this interesting target for further biological investigations.

Very recently, the same group also applied this strategy to the successful total synthesis of (+)-yohimbine (**447**) (*385*). Yohimbine is a well-known indole alkaloid for which the first total synthesis dates back to the 1950s (*386*). Yohimbine has been isolated from several natural sources such as the psychoactive plants *Pausinystalia yohimbe* and *Rauvolfia serpentina* and shows strong α_2-adrenoreceptor antagonist activity (*387*). *Hiemstra's* synthesis of this challenging target employed the phosphoric acid **443**-catalyzed *Pictet-Spengler* reaction of aldehyde **448** and tryptamine derivative **449** to install the tertiary stereogenic center in 84% *ee* (Scheme 99). Intermediate **450** was then converted easily into the *Diels-Alder* synthon **451**, which gave access to yohimbine (**447**) by analogy to a recent report by the *Jacobsen* group (*388*). This thiourea-catalyzed approach will be discussed in Sect. 7.3.

Vinylogous aldol-type reactions have become increasingly more important over the last few years and a variety of approaches have been reported thus far (*389*). Chiral *Brønsted* acid catalysis was applied recently by *Schneider et al.* in vinylogous *Mannich* reactions of preformed silicon dienolates (*390–392*). The versatility of this innovative approach was demonstrated by an impressively short and efficient synthesis of the tobacco piperidine alkaloid (−)-anabasine (**452**) (*392*). Anabasine (**452**) is a minor constituent of the tobacco plant *Nicotiana tabacum*. Using the carefully optimized phosphoric acid catalyst **453**, vinylogous addition of dienolate **454** to imine **455** was carried out in a highly stereocontrolled fashion, giving access to the protected amine **456** in high enantiopurity and almost

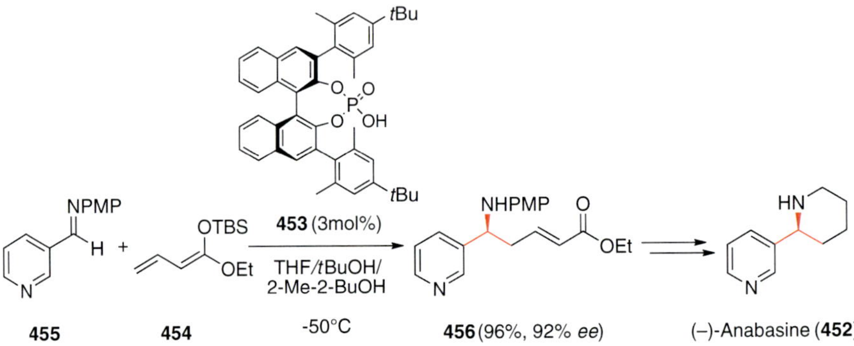

Scheme 100 Chiral acid-catalyzed vinylogous *Mannich* reaction

quantitative yield (Scheme 100). After installation of the required absolute configuration, synthesis of the target **452** then required standard manipulations, resulting in a short four-step synthesis with a high overall yield of 55% (*392*).

7.2 Chiral Diols

In 2003, *Rawal* reported the use of TADDOLs (tetraaryl-1,3-dioxolan-4, 5-dimethanol) as chiral H-bonding catalysts to facilitate highly enantioselective hetero-*Diels-Alder* reactions (*393*). Not surprisingly, this impressive protocol has soon found its way into the repertoire of organic chemists interested in natural product synthesis (*394*, *395*).

In 2004, *Ding et al.* used the naphthyl-based TADDOL **457** to access (*S*)-dihydrokawain (**458**) in a single step from *Brassard's* diene (**459**) and phenylpropanal (**460**) (Scheme 101) (*394*). Although only modest in yield and

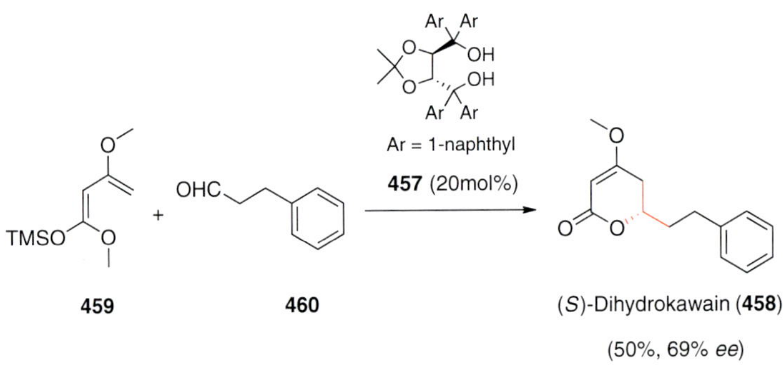

Scheme 101 TADDOL-catalyzed synthesis of (+)-(*S*)-dihydrokawain (**458**)

enantioselectivity (69% *ee*), this report underscores the considerable potential of hydrogen bonding catalysis to obtain chiral naturally occurring motifs in a straightforward way. (A previous synthesis of **458** involving a transition metal-catalyzed hydrogenation to install the stereogenic center required five steps from a commercially available starting material (*396*)). (+)-(*S*)-Dihydrokawain (**458**) is a member of a class of (dihydro)-α-pyrones found in the tropical plant *Piper methysticum*. Beverages made from this plant are an integral part of traditional ceremonies on some Pacific islands and the narcotic, sedative, anti-convulsive, and antifungal properties of **458** makes it an interesting target for the pharmaceutical industry (*397*), thus explaining the interest in novel synthesis strategies like that developed by *Ding et al*.

The complex rocaglate silvestrol (**461**) was isolated from the plant *Aglaia foveolata* by *Kinghorn et al*. (*398–400*). Silvestrol (**461**) shows very potent cytotoxic activity against human lung cancer cells (ED_{50} = 1.2 n*M*), comparable to the activity of the prominent anticancer agent Taxol. In addition, mechanism of action studies indicate that cytotoxicity induced by **461** for human prostate cancer (LNCaP) cells is associated mainly with a block in the cell cycle at the G_2/M checkpoint (*400*, *401*).

Besides the interest of the medicinal chemistry community in this promising biologically active compound, the demanding and complex structural architecture of **461** has attracted the interest of synthesis-oriented organic chemists (Scheme 102).

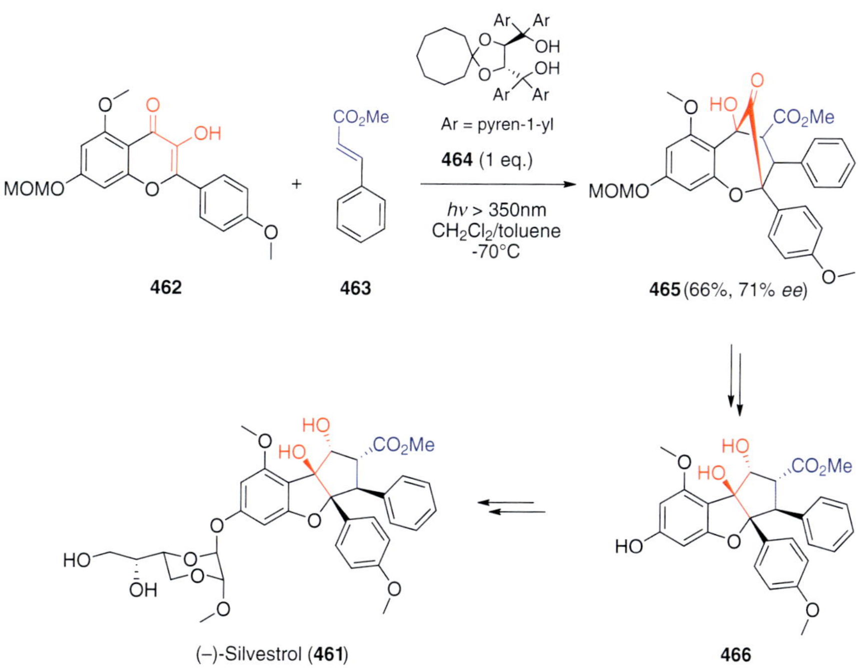

Scheme 102 TADDOL-mediated photocyclization in the total synthesis of silvestrol (**461**)

A first total synthesis of this compound was reported in 2007 by the group of *John A. Porco, Jr.* (*395*). A key step in this synthesis was a photocyclization following a procedure developed earlier by the same group (*402*, *403*) between the hydroxyflavone **462** and cinnamate **463** in the presence of an equimolar amount of TADDOL **464**. Although not catalytic in the use of the chiral hydrogen bonding donor, this example illustrates the potential of chiral diols to facilitate complex transformations in a stereoselective manner. It is worth noting that, in this case, both the nature of the ketal side chain and of the aryl group of the TADDOL are crucial to obtain the cycloadduct **465** in reasonable enantioselectivity (Scheme 102). An elegant α-ketol rearrangement and further functional group manipulations then gave the hydroxyphenyl rocaglate derivative **466** (*395*, *403*), which was finally successfully employed to achieve one of the first total syntheses reported for silvestrol (**461**).

7.3 Chiral (Thio)-Ureas

Chiral (thio)-ureas are possibly the most prominent class of hydrogen bonding catalysts (*366*, *367*). Although both ureas and thioureas have proven their utility through numerous applications, it is fair to say that, of these, thioureas are by far the most commonly used, especially in the syntheses of complex (natural) products. Over the years some impressive examples of their successful application in natural product synthesis have been reported. However, with respect to the exact activation mode, some caution is necessary. Very often, these chiral thiourea catalysts contain an additional catalytically active motif (bifunctional catalysts) like a basic nitrogen or an additional hydrogen bonding donor and sometimes it is not that obvious whether a proper bifunctional activation mode is predominant or not. The following examples depicted in this chapter are believed to be those that are solely hydrogen bonding-mediated, whereas selected examples of bifunctional *Brønsted* acid–base applications of bifunctional thioureas will be covered separately (Sect. 7.4).

The *Jacobsen* group has for some years been among the frontrunners in the development of chiral thioureas and their application in natural product synthesis. In 2007, the highly enantioselective synthesis of indolizidinones and quinolizidinones *via* a thiourea-catalyzed *Pictet–Spengler*-type reaction was reported (*404*). The potential of this protocol may be demonstrated *via* the short four-step total synthesis of (+)-harmicine (**467**). (+)-Harmicine (**467**) is an indole alkaloid isolated from *Kopsia griffithii* by *Kam* and *Sim* (*405*). A leaf extract of this Malaysian plant showed strong antileishmanial activity *in vitro* that was traced to a chromatographic fraction containing (+)-harmicine (**467**) (*405*). The synthesis of **467** was carried out starting from tryptamine (**468**), which, upon treatment with succinic anhydride followed by reduction, was converted initially into hydroxylactam **469**. This compound then underwent an enantioselective intramolecular cyclization in the presence of thiourea catalyst **470** to furnish lactam intermediate **471**. Mechanistic studies suggest that this cyclization proceeds through an S_N1-type pathway *via* the

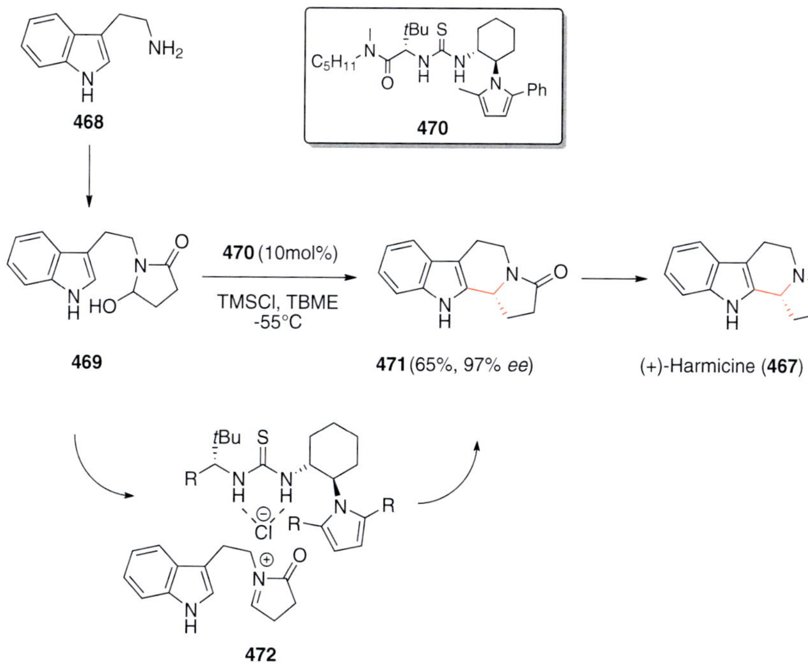

Scheme 103 Thiourea-catalyzed *Pictet-Spengler*-type cyclization in the total synthesis of (+)-harmicine (**467**)

chiral ion pair **472** (Scheme 103). Final lactam reduction then gave the almost enantiopure natural product **467** in good overall yield (*404*).

The same protocol was also applied successfully to the total synthesis of (+)-yohimbine (**447**) (*388*) (A chiral phosphoric acid-catalyzed approach towards this target was discussed earlier (see Scheme 99)). In contrast to *Hiemstra's* approach employing a later stage organocatalytic *Pictet-Spengler* reaction (*385*), *Jacobsen's* synthesis of **447** commenced with the thiourea **473**-catalyzed acyl-*Pictet-Spengler* reaction between tryptamine **468** and the protected hydroxyaldehyde **474**. Intermediate **475** was then employed successfully to access (+)-yohimbine (**447**) in a total of 11 steps and in 14% overall yield (Scheme 104) (*388*).

The tetrahydroisoquinoline ring system is an important structural motif that is commonly encountered in naturally occurring alkaloids with interesting biological activities and has therefore attracted the attention of synthesis-oriented chemists for the last several decades (*406*). The group of *Itoh* investigated the applicability of an asymmetric *Strecker* reaction to synthesize the representative tetrahydroisoquinoline alkaloids (−)-calycotomine (**476**), (−)-salsolidine (**477**), and (−)-carnegine (**478**) (*407*) (Scheme 105). Introduction of the required absolute configuration of these three compounds could be achieved *via* a hydrogen bonding assisted *Strecker* reaction of imine **479** with HCN under cryogenic conditions. A screening procedure for different catalysts identified the *Jacobsen* thiourea catalyst

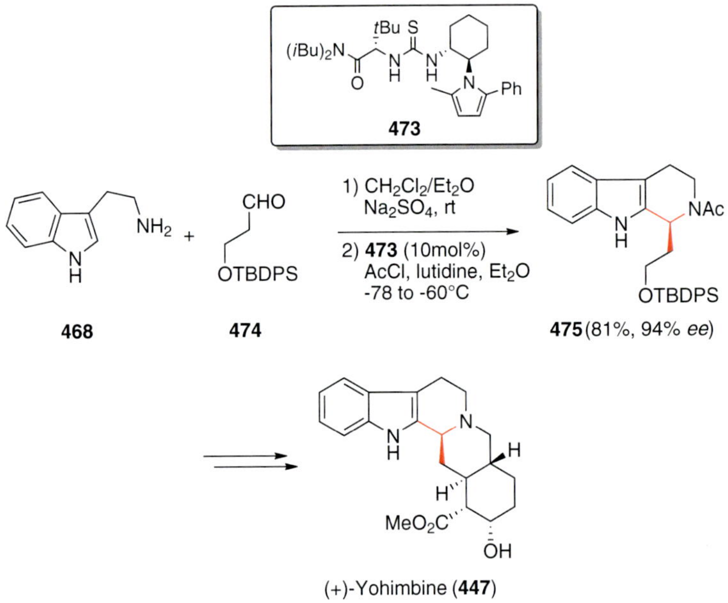

Scheme 104 Thiourea-catalyzed *Pictet-Spengler*-type cyclization in the total synthesis of (+)-yohimbine (**447**)

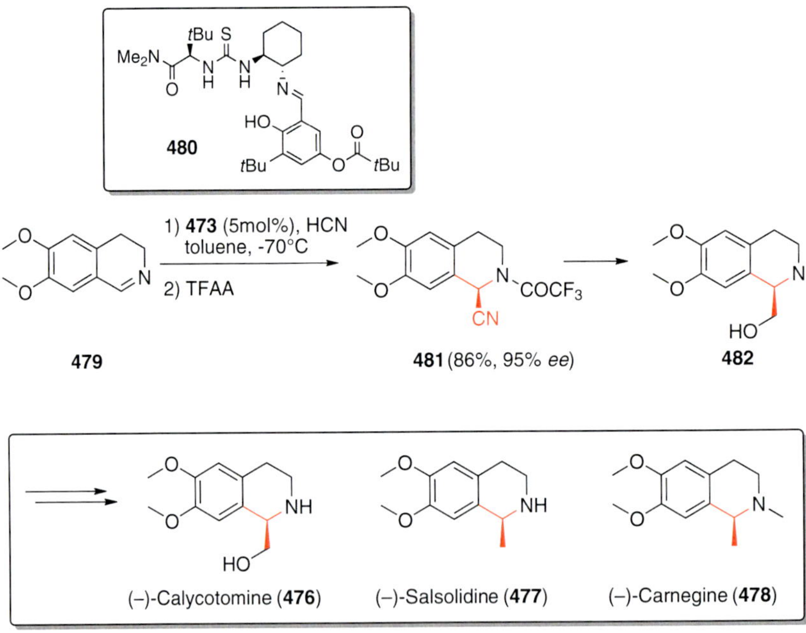

Scheme 105 Thiourea-catalyzed *Strecker* reaction in the syntheses of (−)-calycotomine (**476**), (−)-salsolidine (**477**), and (−)-carnegine (**478**)

480 (*408*) as the one best-suited for this particular transformation, giving the key intermediate **481** in high yield and enantioselectivity (Scheme 105) (*407*). According to mechanistic studies by *Jacobsen et al.*, this catalyst activates the imine *via* hydrogen bonding with both protons of the thiourea functionality (no support for further H-bonding due to the phenolic group was given in the original report) (*408*).

The intermediate **481** was then transformed readily into the tetrahydroisoquinoline alkaloids (−)-calycotomine (**476**), (−)-salsolidine (**477**), and (−)-carnegine (**478**) *via* the hydroxymethyl isoquinoline **482** (*407*).

7.4 Bifunctional *Brønsted* Acid–Base Active (Thio)-Ureas

It was mentioned previously that hydrogen bonding catalysts may contain a second catalytically active motif and the interplay between two different catalytic modes is very often the key to success (bifunctional catalysis). One important example has been depicted in Scheme 4 in the proline-catalyzed aldol reaction. In this case, an enamine activation of the nucleophile and hydrogen bonding of the catalyst to the electrophile resulted in a highly ordered transition state. Also, in the case of the phase-transfer catalyzed asymmetric α-alkylation developed by scientists at Merck Research Laboratories depicted in Scheme 84, the hydroxy group of catalyst **371** was considered to play a pivotal role in this reaction (*19*). Thus, the potential of such bifunctional organocatalysts has been confirmed in several applications. It is fair to say that in these cases a strict separation or classification according to activation mode is not always that obvious. Therefore, some of these examples have already been covered in previous chapters of this volume.

The bifunctional thioureas represent a group of bifunctional catalysts based on an interplay of *Brønsted* acid–base activation that have obtained a prominent position over the last few years. Using these, the second functionality of choice is most often a basic nitrogen, such as a tertiary amine, and a simultaneous activation of nucleophile and electrophile can be achieved (*409*).

Over the last few years, the *Takemoto* group has investigated the applicability of bifunctional thioureas, particularly for *Michael* additions (*410–413*) (Scheme 106).

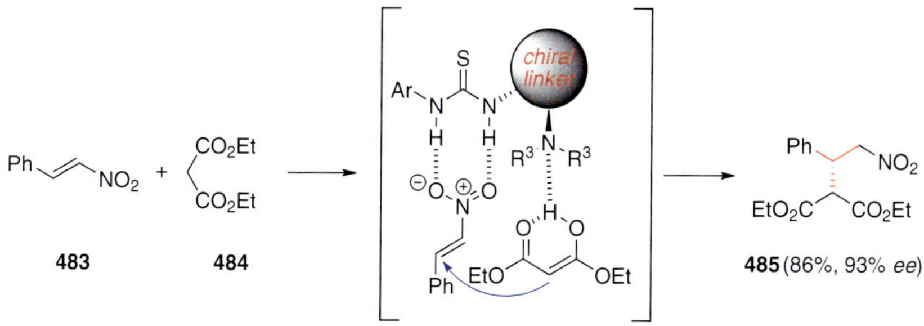

Scheme 106 Bifunctional thiourea-catalyzed *Michael* addition

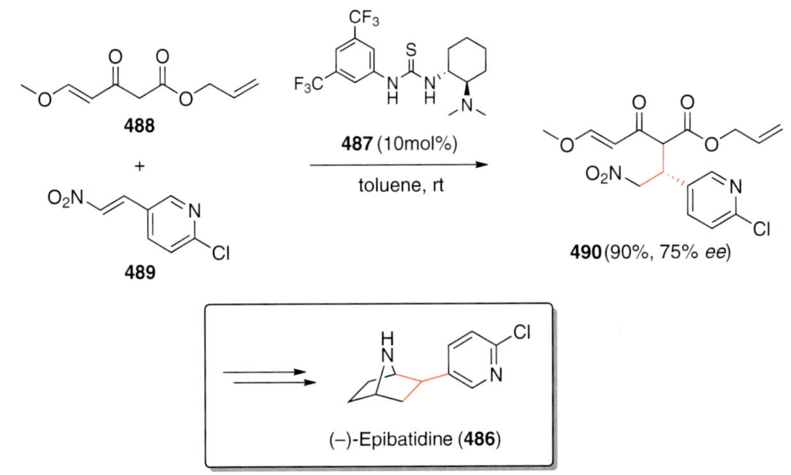

Scheme 107 Bifunctional thiourea **487**-catalyzed *Michael* addition in the total synthesis of (–)-epibatidine (**486**)

The commonly accepted bifunctional activation mechanism of these thioureas containing an additional tertiary amino group involves activation of the electrophile *via* the thiourea group, accompanied by a simultaneous nucleophile activation *via* the amino group in a highly ordered and face-selective fashion due to the chiral linker (scaffold).

An early application of this strategy to access a chiral natural product was reported by the *Takemoto* group in 2006 in their total synthesis of (–)-epibatidine (**486**) (*413*). This alkaloid was isolated from the skin of the Ecuadorian frog *Epipedobates tricolor* in the early 1990s (*414*) and has been found to possess non-opiate analgesic properties around 200 times more potent than morphine (*415*). Synthesis of this interesting target commenced with a bifunctional thiourea **487**-catalyzed *Michael* addition of β-keto ester **488** to acceptor **489** (Scheme 107). Surprisingly, the enantiomeric excess in this step was lower (75%) than in other test reactions investigated in this report (*413*). Nevertheless, the intermediate **490** was employed successfully to access the alkaloid (–)-epibatidine (**486**) in a few further steps. Carrying out a recrystallization of an advanced intermediate later in the sequence increased the enantiopurity.

The *Barbas* group recently employed the same catalyst in 1,4-additions of oxindoles to nitroalkanes, a concept that was also employed successfully to a synthesis of (+)-esermethole (**377**) (*416*). As discussed earlier (Sect. 6.1 – Scheme 86), esermethole is an important precursor of the naturally occurring anticholinesterase agent physostigmine (**378**) (*341*). Addition of oxindole **491** to nitroalkene **492** catalyzed by **487** gave the (+)-esermethole precursor **493** in reasonable enantiopurity (83% *ee*). After recrystallization, the enantiomeric excess could be enhanced to 96% and **493** was then transformed into (+)-esermethole (**377**) in two further steps and an impressive overall yield of 72% (Scheme 108).

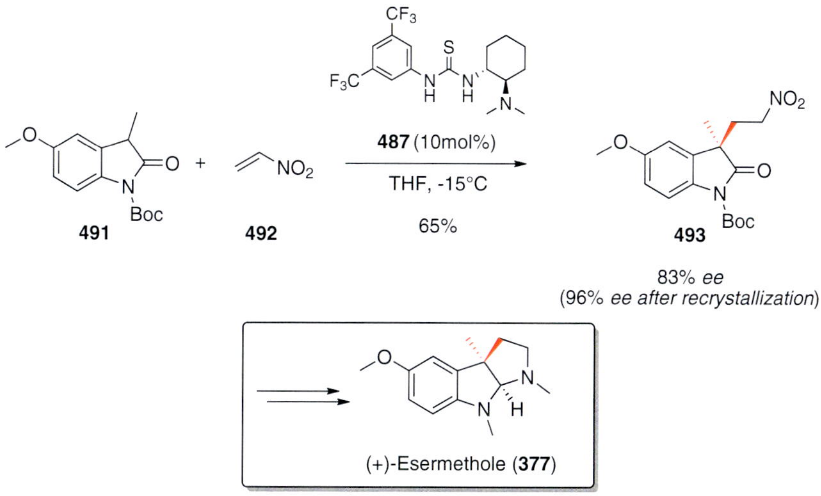

Scheme 108 *Michael* addition in the synthesis of (+)-esermethole (**377**)

Besides the use of bifunctional thioureas containing a chiral cyclohexane-type linker like catalyst **487**, the use of *Cinchona* alkaloid-derived (thio)-ureas has found to be highly popular and fruitful (*366*, *367*, *417–420*). In these cases, the easily obtainable well-defined spatial arrangement of the naturally occurring *Cinchona* alkaloids can be exploited efficiently to build up highly selective and versatile bifunctional organocatalysts.

In 2008, *Falck* and co-workers reported an enantioselective organocatalytic oxa-*Michael* addition to prochiral enones using boronates as hydroxide nucleophile equivalents in the presence of a bifunctional H-bonding catalyst (*421*). The potential of this methodology was proven in a short synthesis of acetate **494**, a potent antifungal and hepatoprotective compound isolated from avocados (*422*) and in the synthesis of (+)-(*S*)-streptenol A (**495**), one of four known streptenols produced by *Streptomyces luteogriseus* that has attracted attention as an immunostimulant as well as an inhibitor of cholesterol biosynthesis and tumor cell proliferation (*423*). The key steps in both syntheses were *Cinchona* alkaloid **496**-catalyzed oxa-*Michael* additions of phenylboronic acids (**497** or **498**) to enones (**499** or **500**). Subsequent oxidation gave the corresponding alcohols in high yields and enantioselectivities (Scheme 109). Whereas in the case of enone **499** phenylboronic acid **497** worked well, enone **500** required a more efficacious nucleophilic partner to obtain the corresponding aliphatic diol with a reasonable reaction rate (*421*).

For some years the group of *Li Deng* has been investigating carefully the use of *Cinchona* alkaloids in asymmetric organocatalysis resulting in the development of numerous highly useful applications (*424*). Recently, they developed a catalytic tandem conjugate addition/protonation protocol to access compounds with tertiary

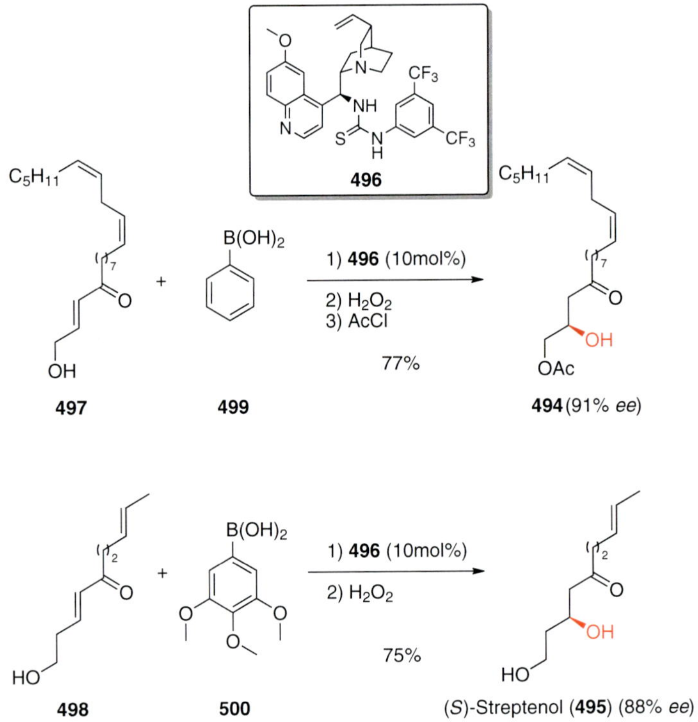

Scheme 109 Bifunctional *Cinchona* alkaloid-catalyzed oxa-*Michael* additions

and quaternary stereogenic centers in a 1,3-relationship with good enantio- and diastereoselectivity (*425*, *426*). By exploring different catalysts for this transformation, it was found that bifunctional catalysts (like **501**) exhibit complementary diastereoselectivities with respect to non-bifunctional ones (*426*). Impressively, this methodology allowed for the first time the stereoselective construction of 1,3-related tertiary and quaternary stereogenic centers in either of the possible configurations by starting from the same starting materials. The potential for complex natural product synthesis was proven by its successful application to the asymmetric total synthesis of manzacidin C (**502**). Manzacidin C belongs to a family of bromo-tetrahydropyrimidine alkaloids that were isolated from the Okinawan marine sponge *Hymeniacidon* sp. (*427*), and it has been shown that they possess potential as R-adrenoceptor blockers, antagonists of serotonergic receptors, and actomyosin ATPase activators (*428*). Due to its scarcity when obtained from natural sources, these compounds have attracted considerable interest from the synthesis-oriented chemical community (*429*, *430*). *Li Deng's* approach to access manzacidin C (**502**) employed their newly developed conjugate addition of **503** to *Michael* acceptor **504** catalyzed by the bifunctional *Cinchona* alkaloid catalyst **501**. Compound **505** could thus be obtained almost quantitatively

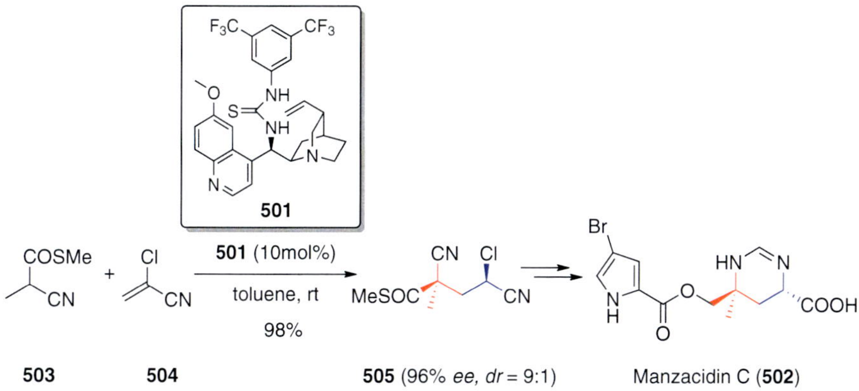

Scheme 110 Bifunctional catalysis in the synthesis of manzacidin C (**502**)

and with perfect selectivity and was then employed successfully to access manzacidin C (**502**) in 12 additional steps (Scheme 110).

The group of *Dixon* has reported the use of *Cinchona* alkaloid-derived bifunctional H-bonding catalysts for the syntheses of drugs like rolipram or paroxetine (*420*) as well as for the naturally occurring marine alkaloid (−)-nakadomarin (**506**) (*417*) (Scheme 111). Nakadomarin (**506**) was isolated from the sponge

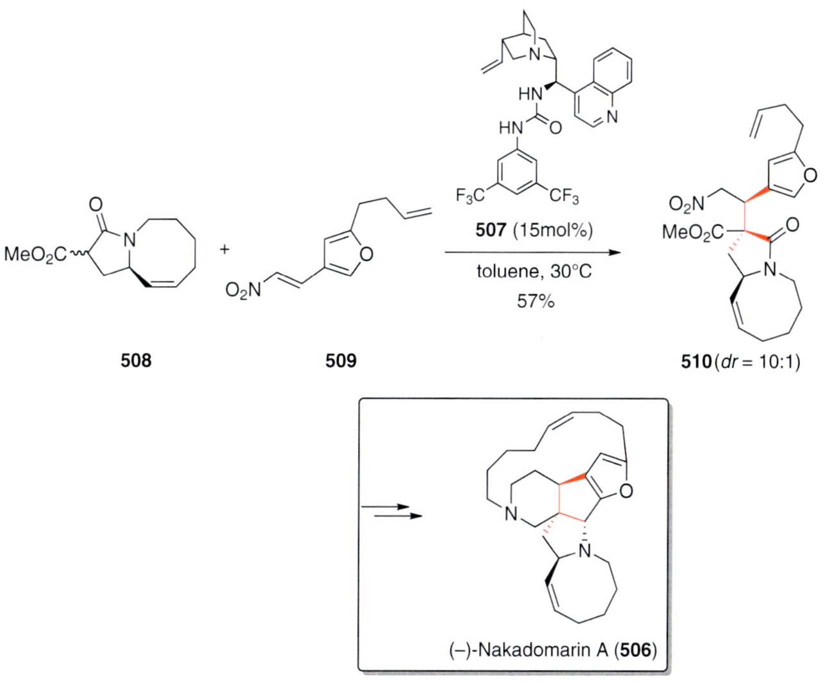

Scheme 111 *Cinchona* alkaloid-derived bifunctional urea catalyst **507** in the total synthesis of (−)-nakadomarin (**506**)

Amphimedon sp. off the coast of the Kerama islands, Okinawa, Japan and showed cytotoxicity against L1210 murine lymphoma cells (*431*, *432*). One of the key steps in the highly impressive synthesis of this challenging target by *Dixon* was the use of the bifunctional urea-based *Cinchona* catalyst **507** to facilitate the intramolecular *Michael* addition between the chiral lactam **508** and the nitro olefin **509**. Employing this strategy, the targeted diastereomer **510** could be isolated in reasonable yield (57%) and with good diastereoselectivity (10:1) (Scheme 111). With all the stereogenic centers built up in the required configuration for the targeted endgame, intermediate **510** was then converted into (−)-nakadomarin (**506**) in a series of efficient transformations (*417*). Very recently, the same group reported a slightly modified route to access (−)-nakadomarin (**506**) *via* a late-stage alkyne ring-closing metathesis approach. Using this procedure, the organocatalytic key step remained the same, with slightly adapted synthons utilized (*433*).

The family of the abyssinone natural products has been known for its interesting and promising biological properties for some time (*434*, *435*). Isolated from different plants traditionally used in folk medicines, these compounds show antimicrobial activity (*434*) and aromatase inhibition (*435*). The main challenge in the syntheses of these compounds is a chiral flavanone core. *Scheidt et al.* (Scheme 112)

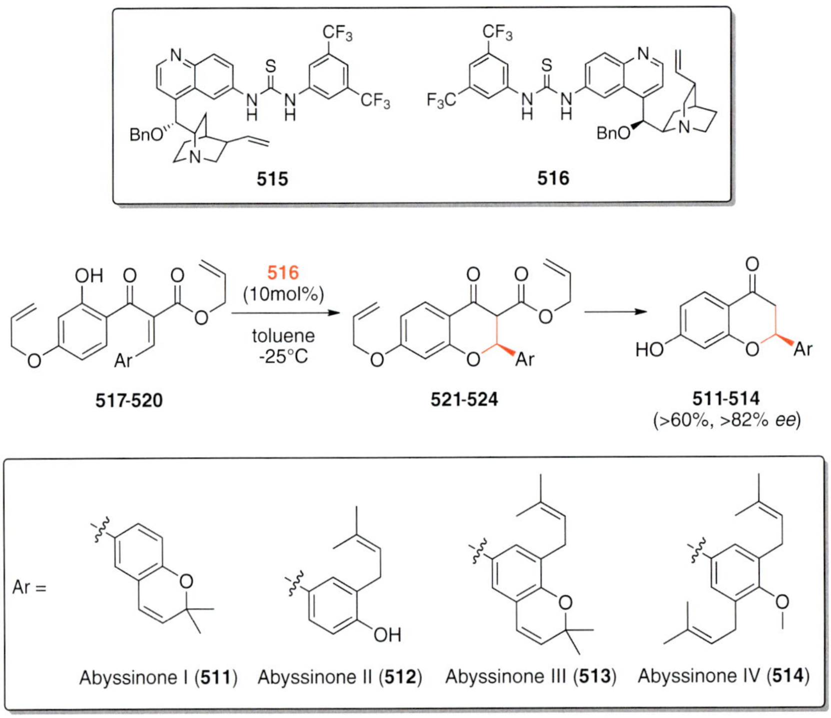

Scheme 112 Thiourea-catalyzed access towards abyssinones I–IV (**511–514**) and their enantiomers (obtained using catalyst **515**)

have recently published a concise approach to synthesize abyssinones I–IV (**511**–**514**) as well as their antipodes *via* a bifunctional thiourea **515**- or **516**-catalyzed intramolecular oxa-*Michael* addition as the key step (*418*). Following this procedure, they were able to install the required configuration at the C-2-position using the easily available appropriately substituted starting materials **517**–**520**. The corresponding 3-carboxyflavanones could thus be obtained in high yields and with good enantioselectivities having the absolute configuration controlled by choice of the correct pseudoenantiomer of the catalyst employed (Scheme 112). A final tandem deprotection/decarboxylation step then furnished the natural (*S*)-configured products abyssinones I–IV (**511**–**514**) and their enantiomers in sufficient quantities for further biological testing (*418*).

Very recently, *Fan* and co-workers employed a thiourea-catalyzed *Michael* addition early in the total syntheses of (−)-lycoramine (**525**), (−)-galanthamine (**526**), and (+)-lunarine (**527**) (*436*). The alkaloids (−)-lycoramine (**525**) and (−)-galanthamine (**526**) belong to the Amaryllidaceae group of alkaloids. Lycoramine was first isolated from the red spider lily *Lycoris radiata* in 1932 (*437*), and galanthamine was originally isolated from the Caucasian snowdrop *Galanthus woronowi* in 1952 (*438*). Since the 1990s, (−)-galanthamine (**526**) has been used clinically as a selective acetylcholinesterase inhibitor for the treatment of *Alzheimer's* disease (*439*), while (−)-lycoramine (**525**) was found to show similar acetylcholinesterase-inhibitory activity and is also claimed to inhibit peptide bond-formation during protein synthesis (*440*). The *Lunaria* alkaloid (+)-lunarine (**527**) was isolated initially from a *Lunaria* species, an ornamental plant, at the beginning of the last century (*441*) and demonstrated inhibition of trypanothione reductase (TryR), an important factor in the defense of certain parasites against reactive oxygen species generated by host cells (*442*).

The structural key feature of these potent alkaloids is a functionalized *cis*-hydrodibenzofuran core with an all-carbon quaternary stereogenic center. The *Fan* group used a thiourea **487**-catalyzed *Michael* addition of compound **528** to acrylate **529** to obtain the lycoramine- and galanthamine-synthon **530** in good yield and with almost perfect enantiopurity after a single recrystallization of the initial product (*ee* increased from 80 to 99%). Synthesis of (+)-lunarine (**527**) required use of the *Cinchona* alkaloid-derived thiourea catalyst **531** to obtain the required *Michael* product **533**. While **530** was then used to synthesize the advanced spirocyclic intermediate **534** giving access to both (−)-lycoramine (**525**) and (−)-galanthamine (**526**), compound **533** was converted into (+)-lunarine (**527**) in a similar manner (Scheme 113) (*436*).

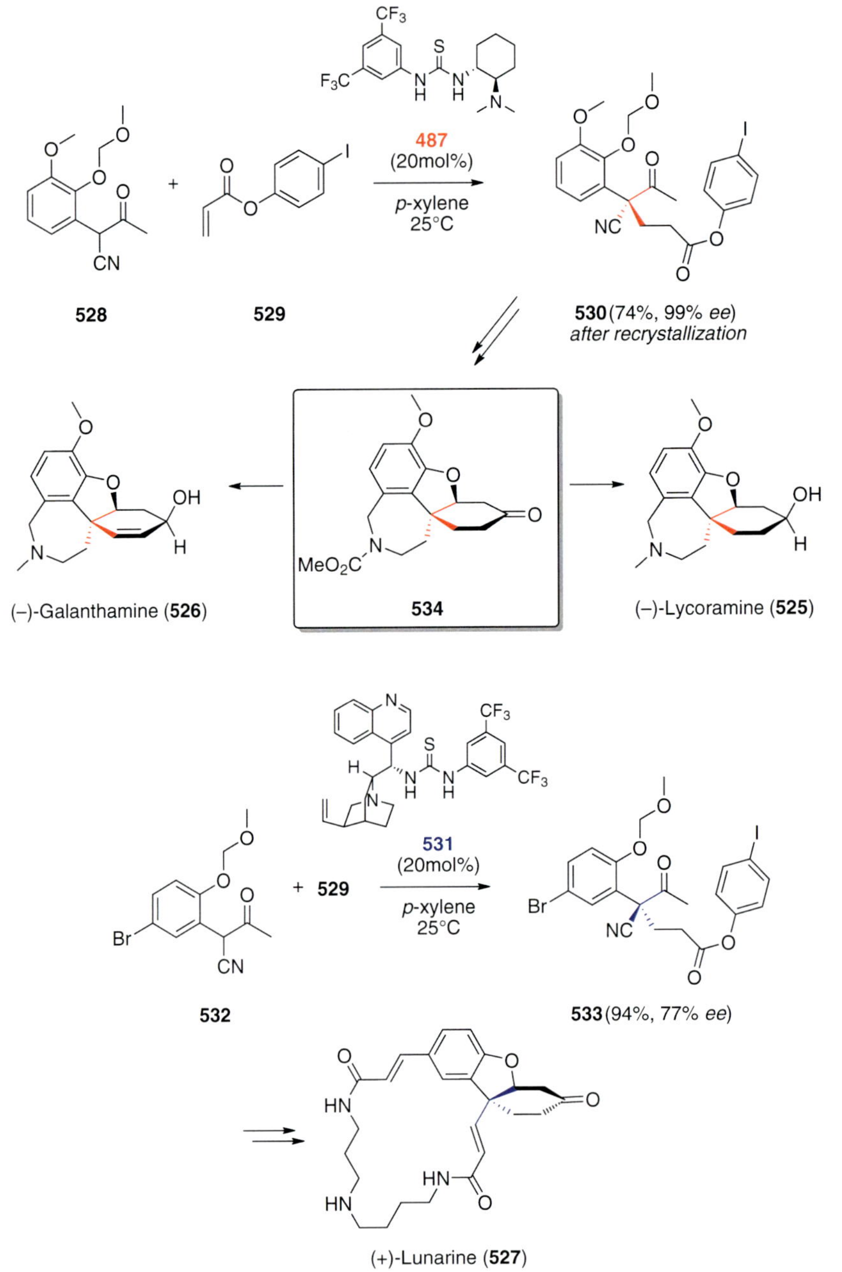

Scheme 113 Organocatalytic transformation in the total syntheses of (−)-lycoramine (**525**), (−)-galanthamine (**526**), and (+)-lunarine (**527**)

7.5 Synopsis

Chiral H-bond donors and acids have proven their potential many times over several decades. Some useful applications in natural product synthesis have been reported, using either hydrogen bonding activation as the sole catalytically active principle, or utilizing bifunctional catalysts. With respect to the catalytic moiety of choice, the considerable potential of thioureas can be emphasized, especially those based on *Cinchona* alkaloids (Table 6).

Table 6 Chiral acids and H-donors in natural product syntheses

Catalyst	Product	References
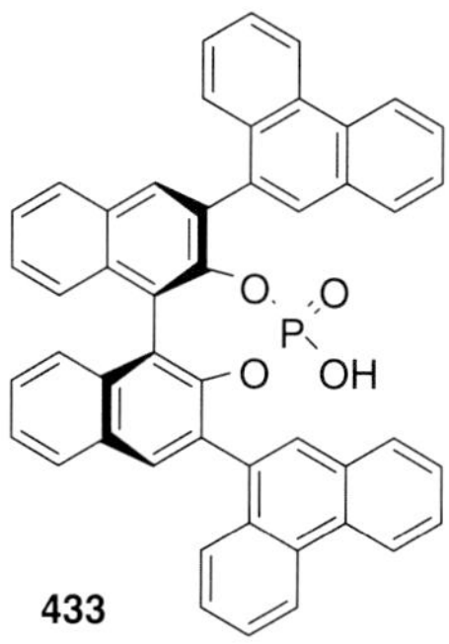 433	(+)-Galipinine (**436**) (+)-Cuspareine (**437**) (−)-Angustureine (**308**)	*(376)* *(376)* *(376)*
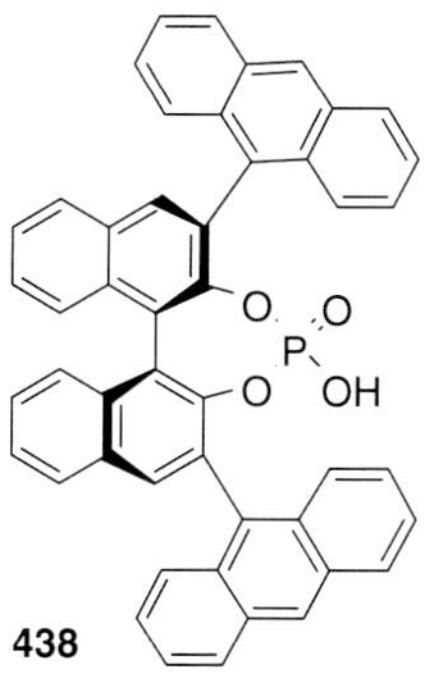 438	*diepi*-Pumiliotoxin C (**441**)	*(379)*
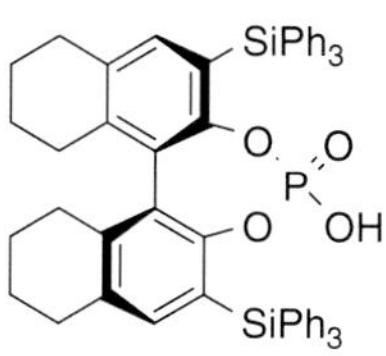 443	Arboricine (**442**) (+)-Yohimbine (**447**)	*(383)* *(385)*

(continued)

Table 6 (continued)

Catalyst	Product	References
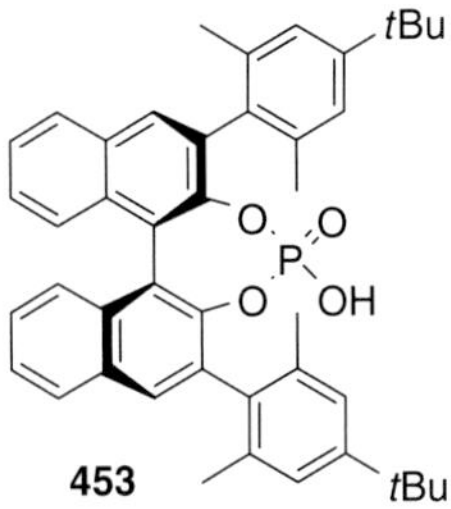 **453**	(−)-Anabasine (**452**)	(*392*)
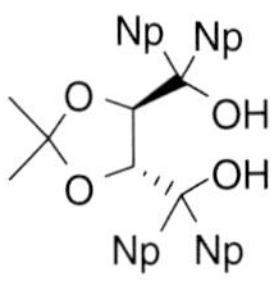 **457**	(*S*)-Dihydrokawain (**458**)	(*394*)
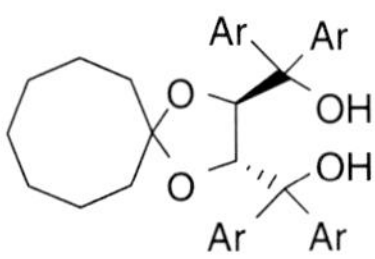 Ar = pyren-1-yl **464**	Silvestrol (**461**)	(*395*)
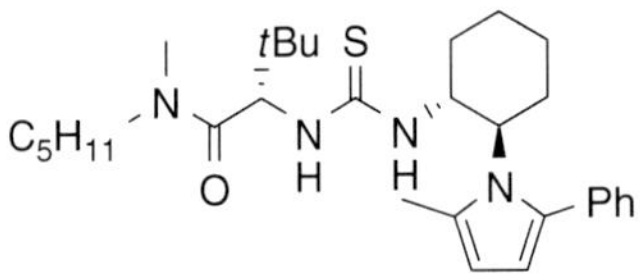 **470**	(+)-Harmicine (**467**)	(*404*)
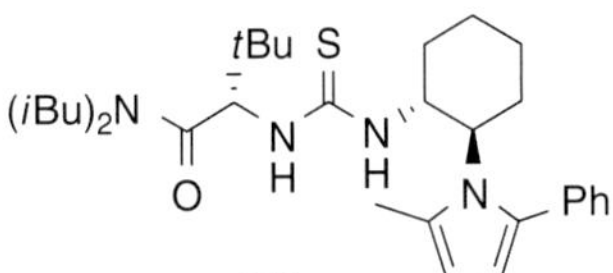 **473**	(+)-Yohimbine (**447**)	(*388*)

(continued)

Table 6 (continued)

Catalyst	Product	References
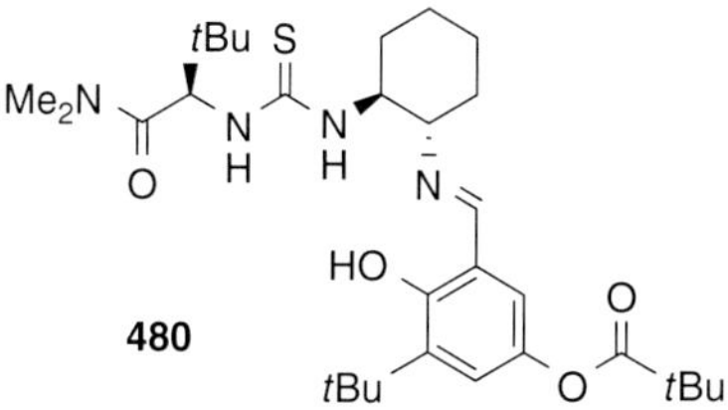 **480**	(−)-Calycotomine (**476**) (−)-Salsolidine (**477**) (−)-Carnegine (**478**)	(*407*) (*407*) (*407*)
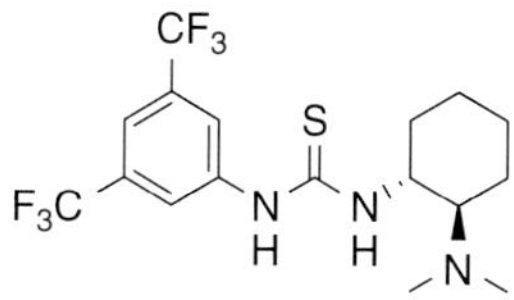 **487**	(−)-Epibatidine (**486**) (+)-Esermethole (**377**) (−)-Lycoramine (**525**) (−)-Galanthamine (**526**)	(*413*) (*416*) (*436*) (*436*)
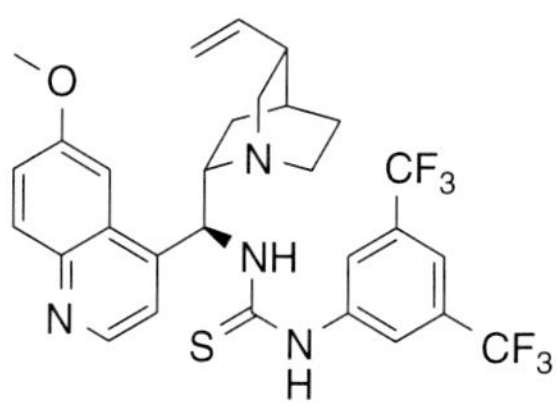 **496**	(+)-(*S*)-Streptenol A (**495**)	(*421*)
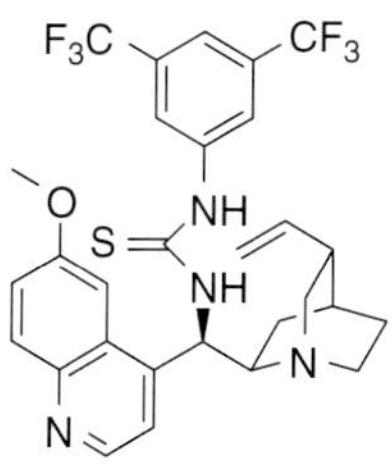 **501**	Manzacidin C (**502**)	(*426*)
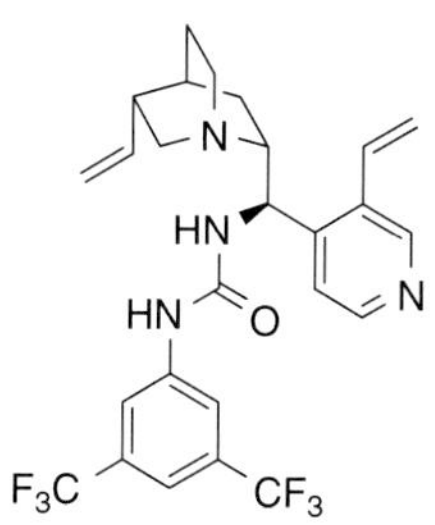 **507**	(−)-Nakadomarin (**506**)	(*417*)

(continued)

Table 6 (continued)

Catalyst	Product	References
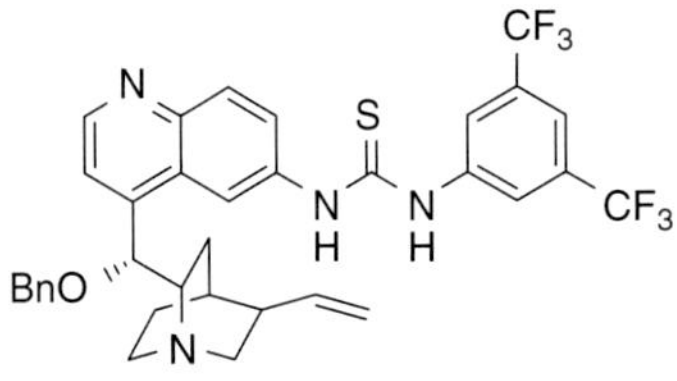 **515**	Abyssinones I–IV (**511–514**)	(*418*)
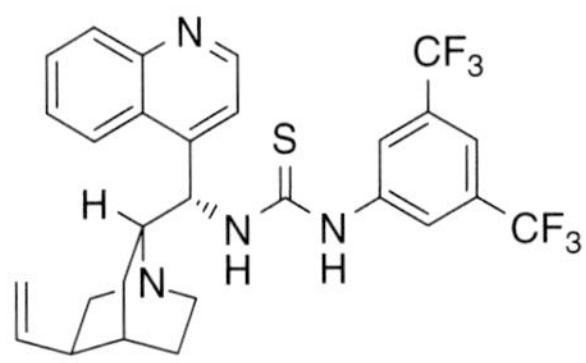 **531**	(+)-Lunarine (**527**)	(*436*)

8 Chiral *Brønsted* and *Lewis* Bases

The catalytic potential of base functionalities has been referred to in the previous chapter (see Sect. 7.4), wherein the interplay between an acidic (thio-)urea and a basic amine separated by a chiral linker was shown to enable the simultaneous activation of both the electrophile and nucleophile. In addition to such bifunctional thiourea-containing acid–base catalysts, chiral catalysts containing (*Lewis* or *Brønsted*-) base functionality as the sole catalytically active group as well as those having another H-bond donor like a hydroxy group (*e.g. Cinchona* alkaloids) have found widespread applications in asymmetric catalysis (*443–449*).

The potential of these catalysts is due to the fact that a variety of different activation modes are possible, thus facilitating their application for different types of reactions. On the one hand, chiral bases can be used to carry out face-selective deprotonations and the formation of chiral ion pairs, but, on the other hand, such compounds can also be used as nucleophilic catalysts. The following sections will be divided primarily according to the class of catalysts employed and to a lesser extent according to the proposed activation mode.

8.1 *Cinchona* Alkaloids

Tertiary amines have found to be among the most useful chiral bases used in modern (asymmetric) catalysis. With respect to the chiral backbone of these catalysts, *Cinchona* alkaloids have emerged as the most commonly employed representatives and have proven their potential in numerous demanding applications (*424*, *443*).

As mentioned in the introductory chapter, historically the first asymmetric organocatalytic reaction can be dated back to *Breding's* quinine (**2**)- or quinidine (**3**)-mediated addition of HCN to benzaldehyde (**1**) at the beginning of the last century (Scheme 1) (*7*), a reaction that was later on reinvestigated by *Prelog* (*15*). In addition, one of the first highly enantioselective reactions ever was reported in the 1950s by *Pracejus*, who carried out the addition of methanol to methyl phenyl

M. Waser, *Asymmetric Organocatalysis in Natural Product Syntheses*,
Progress in the Chemistry of Organic Natural Products 96,
DOI 10.1007/978-3-7091-1163-5_8,

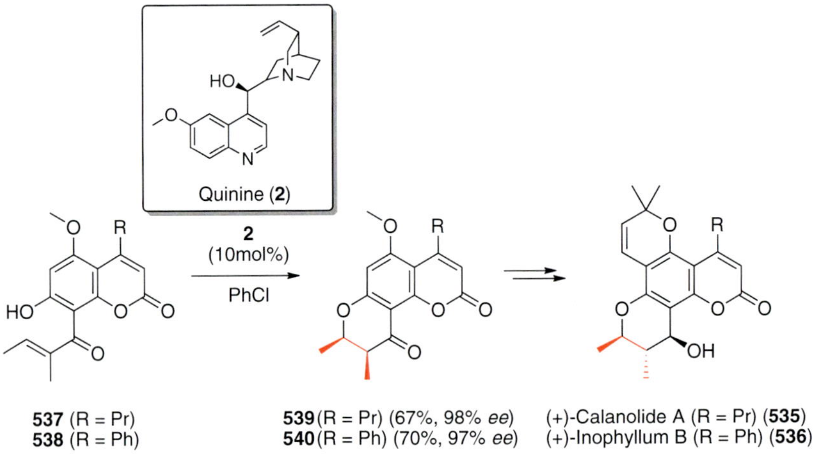

Scheme 114 Quinine-catalyzed oxo-*Michael* addition in the total syntheses of (+)-calanolide A (**535**) and (+)-inophyllum B (**536**)

ketene (**5**) in the presence of *O*-acetylquinine (**6**) (Scheme 2) (*16*). Thus, *Cinchona* alkaloids are among the most important and versatile chiral natural products available with respect to their application in asymmetric catalysis. Besides the use of natural *Cinchona* alkaloids, modified derivatives of these compounds have been used increasingly over the last few years.

In 1981, *Hiemstra* and *Wynberg* reported a thorough investigation of the *Cinchona* alkaloid-catalyzed addition of thiols to α,β-unsaturated enones (*450*). This report may now be considered as one of the major breakthroughs in asymmetric organocatalysis, as it has set the stage for a large number of different applications based on this elegant concept. Mechanistic studies revealed that *Cinchona* alkaloids most likely act as bifunctional catalysts in these reactions (*450*). This methodology was extended later to the use of oxygen nucleophiles, as demonstrated by the total syntheses of (+)-calanolide A (**535**) and (+)-inophyllum B (**536**) by *Ishikawa et al.* (*451*). (+)-Calanolide A (**535**) is a potent anti-HIV-1 inhibitory coumarin derivative isolated in 1992 from the tropical evergreen tree *Calophyllum lanigerum* var. *austrocoriaceum* (*452*). (+)-Inophyllum B (**536**) was isolated approximately at the same time from *Calophyllum inophyllum* and proved to be an active inhibitor of HIV-reverse transcriptase (*453*). In their syntheses of these potent agents, *Ishikawa et al.* employed a quinine (**2**)-catalyzed intramolecular oxo-*Michael* addition to install the stereogenic centers of the chromanol ring. Carrying out this reaction on precursors **537** and **538**, the corresponding chromanone skeleton could be constructed in high enantioselectivity (>97% *ee*) and acceptable diastereoselectivity (>67% yield of the *cis*-isomers) (Scheme 114). The functionalized chromanones **539** and **540** were used to yield (+)-calanolide A (**535**) and (+)-inophyllum B (**536**) in only three more steps, including a required isomerization of one stereogenic center in order to obtain the targeted *trans*-substitution of the natural products (*451*).

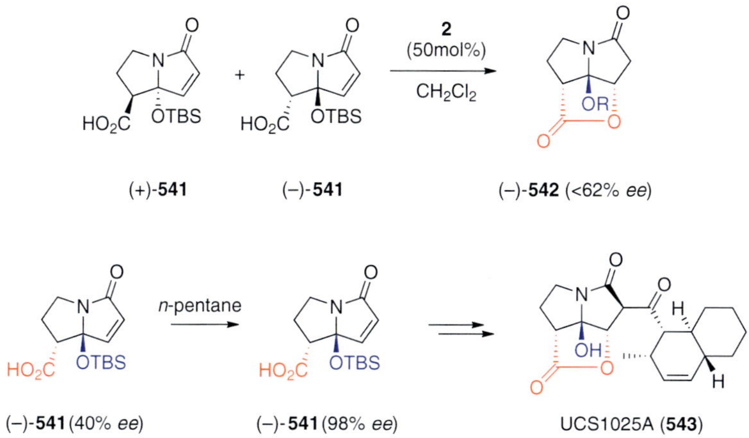

Scheme 115 *Christmann's* quinine-mediated kinetic resolution to access carboxylic acid (−)-**541**, a key intermediate to access UCS1025A (**543**)

The *Christmann* group used quinine (**2**) for the kinetic resolution of racemic pyrrolizidine **541** (*454*). They found that stirring a solution of **541** in dichloromethane in the presence of 50 mol% **2** resulted in the formation of the enantioenriched lactone (−)-**542**. Notably, the carboxylic acid (−)-**541** is a key synthon in *Danishefsky's* total synthesis of UCS1025A (**543**) (*455*), a compound isolated from the fermentation broth of the *Acremonium* sp. KY4917 fungus, and shown to possess antiproliferative activity against human cancer cell lines by inhibition of the telomerase enzyme (*456*, *457*). To obtain (−)-**541** in sufficient enantiopurity as a starting material for the synthesis of **543**, *Christmann et al.* developed a very efficient protocol by triturating enantioenriched (−)-**541** (around 40% *ee*, either obtained by carrying out the kinetic resolution in the presence of quinidine (**3**), or by converting back (−)-**542** (*454*)) in hot *n*-pentane. It was found that (−)-**541** is readily dissolved whereas racemic **541** remained undissolved. Following this facile and highly efficient protocol, (−)-**541** could be obtained almost in enantiopure form and in gram quantities (Scheme 115). Synthesis of UCS1025A (**543**) could then be accomplished by following *Danishefsky's* route (*455*).

The group of *Wang* recently investigated the total synthesis of (−)-spirobrassinin (**544**) (*458*), a natural product isolated from *Pseudomonas cichorii*-inoculated Chinese cabbage and Japanese radishes (*459*). Interestingly, this compound displays various biological properties like plant defense and antifungal activities (*460*, *461*) as well as potential cancer chemopreventive activity (*462*). *Wang's* elegant strategy to access the spiro-stereogenic center of this target involves a highly stereoselective nitroaldol reaction (*Henry* reaction) of nitromethane (**278**) to isatin (**545**) in the presence of cupreine (**546**) and benzoic acid as an additive in dimethylacetamide (DMA). Of significance, **546** was found to be a better catalyst for this transformation than other bifunctional catalysts tested such as

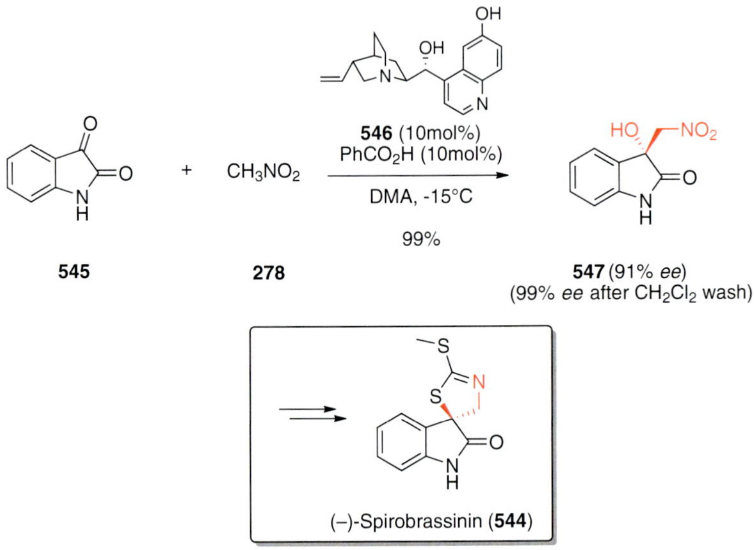

Scheme 116 Cupreine (**546**)-catalyzed *Henry* reaction in the total synthesis of (−)-spirobrassinin (**544**)

thiourea-containing *Cinchona* alkaloids (*458*). Using this operationally simple procedure, the key intermediate **547** was obtained quantitatively and in high enantiomeric excess (washing the crude product with CH_2Cl_2 resulted in further *ee* improvement). With compound **546** in hand, the successful total synthesis of (−)-spirobrassinin (**544**) required just two additional steps (Scheme 116).

The chiral ketene dimers (*R*)- and (*S*)-**548** are valuable building blocks for the synthesis of polyketides (*463–466*). A useful approach to obtain these compounds in high enantiopurity is the treatment of methylketene (**549**) with catalytic amounts of either TMS-protected quinine (**550**) (**2** gives (*S*)-**548** in lower *ee* only) or quinidine (**3**) (*467*). Mechanistically, this transformation provides an excellent example of the high catalytic potential of *Cinchona* alkaloids as nucleophilic catalysts (Scheme 117). The *Calter* group used both enantiomers of **548** as starting materials to access different polyketide natural products with various interesting biological properties like siphonarienolone (**551**), siphonarienedione (**552**), or siphonarienal (**553**) (*464*, *466*), as well as a key segment of the potent antiobiotic pamamycin 621A (**554**) (*463*). As the organocatalytic key reaction step in these approaches is the synthesis of the starting material **548**, a detailed discussion of these total syntheses seems to be beyond the scope of this volume, and the interested reader is referred to the original work of *Calter et al.* (*463–466*).

The potential of *Cinchona* alkaloids as nucleophilic catalysts was also demonstrated in *Gaunt's* cyclopropanation approach by reacting α-halo carbonyl compounds with *Michael* acceptors in the presence of catalytic amounts of O-protected *Cinchona* alkaloids (*468–470*). This reaction is thought to proceed

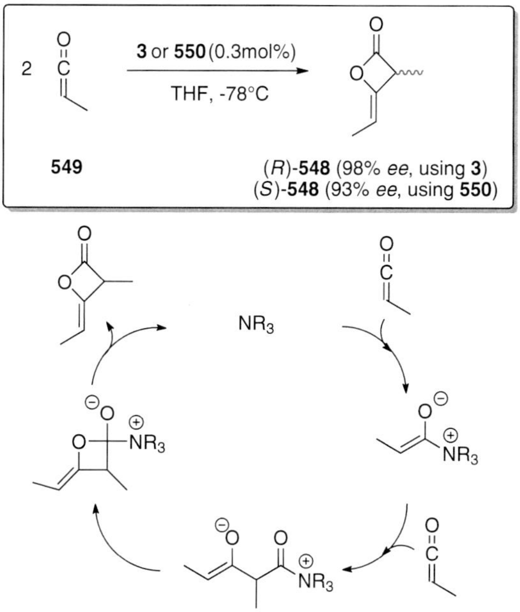

Scheme 117 *Cinchona* alkaloid-catalyzed dimerization of methylketene (**549**)

via an *in situ*-formed ammonium ylide. Of note, while chiral ammonium ylides allow the syntheses of different cyclopropanes in a stereoselective way in high yields, chiral epoxide formation using such a strategy has failed so far (*471*, *472*).

It comes as no surprise that this useful methodology has also found an application in natural product synthesis (*473*, *474*). The *trans*-cyclopropane motif is a prevalent structural feature of some members of the oxylipin family (*475*). *Kumaraswamy* and co-workers have recently described the synthesis of an eicosanoid (**555**) (*473*) and a lactonealdehyde (**556**) (*474*). While **555** is believed to be an important intermediate in the biogenetic route towards different members of the oxylipin family (*473*), **556** is an advanced synthetic intermediate towards solandelactones A and B (**557** and **558**) (*474*), which were isolated in 1996 from the hydroid *Solanderia secuda*, collected off the Korean coast (*476*).

Synthesis of eicosanoid **555** commenced with a stereoselective organocatalytic cyclopropanation achieved by reacting bromoacetate **559** with *Michael* acceptor **560** in the presence of dimeric catalyst **561**. As the authors were not able to resolve product **562** using an enantioselective HPLC phase, the exact enantiomeric excess could not be determined on this stage but was found to be satisfactory for the rest of the sequence. With respect to the synthesis of lactonealdehyde **556**, the cyclopropanation was carried out by adding the *Weinreb*-amide **563** to enone **564** to furnish the key intermediate **565**, which was transferred further into **556** in a similar manner (*474*) (Scheme 118).

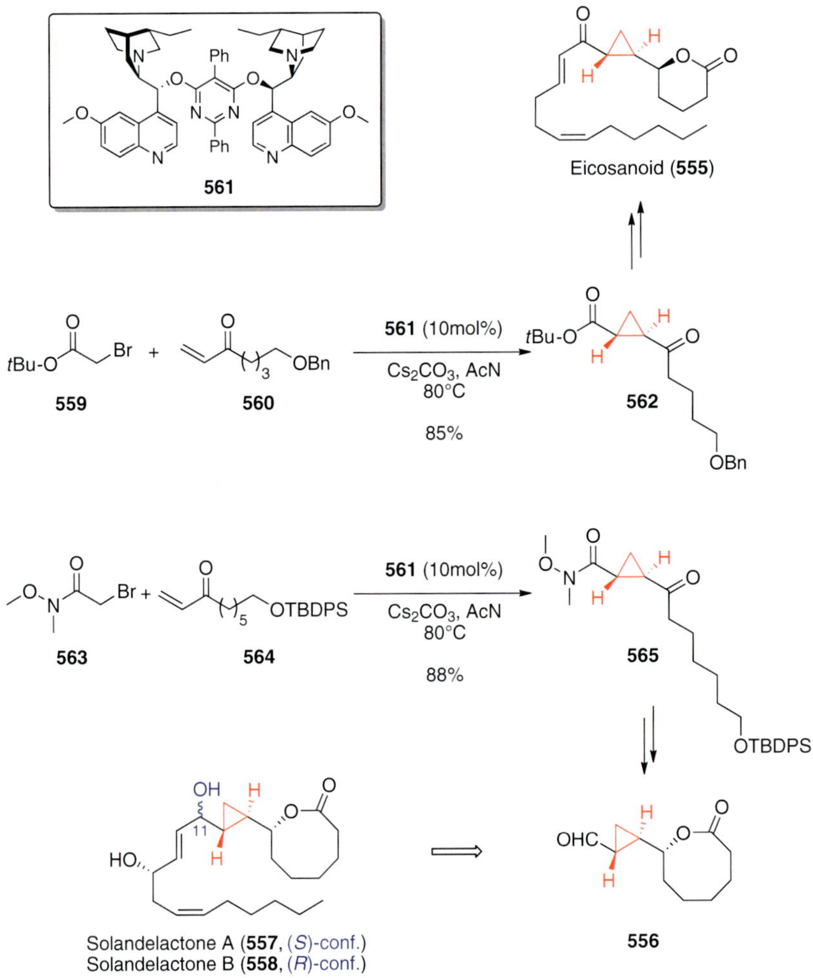

Scheme 118 *Cinchona* alkaloid-catalyzed cyclopropanation in the synthesis of members of the oxylipin family

Another class of reaction for which chiral tertiary amines are privileged catalysts is the *Morita-Baylis-Hillman* type (*477*, *478*). One of the first applications of *Cinchona* alkaloids to mediate an asymmetric *Morita-Baylis-Hillman* reaction in a natural product synthesis was reported by *Hatakeyama et al.* in 2001 (*479*). Using a stoichiometric amount of β-isocupreidine (**568**), a stereoselective addition of hexafluoroisopropyl acrylate (**569**) to aldehyde **570** could be carried out in good yield and with excellent selectivity (99% *ee*) (Scheme 119). The chiral β-hydroxy ester **571** was converted further into the epoxide **572**, a known intermediate in the synthesis of epopromycin B (**573**). Epopromycin B (**573**) is a plant cell wall

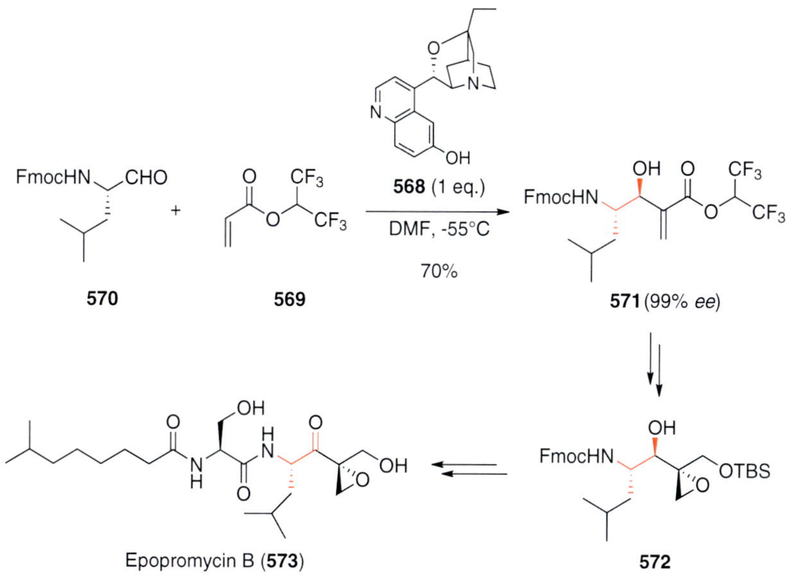

Scheme 119 Asymmetric *Morita-Baylis-Hillman* reaction in the synthesis of a key intermediate towards epopromycin B (**573**)

synthesis inhibitor isolated from the culture broth of *Streptomyces* sp. NK0400 (*480*).

Hatakeyama et al. succeeded shortly afterwards in carrying out this type of reaction in a catalytic fashion, as demonstrated first in the synthesis of the potent immunosuppressant (−)-mycestericin E (**574**) (*481*), followed by a recent formal total syntheses of (+)-fostriecin (**575**) and (+)-phoslactomycin B (**576**), employing a key organocatalytic stereoselective *Morita-Baylis-Hillman* step (*482*). The phoslactomycin family has attracted interest not only for its intriguing structures but also as lead compounds for novel anticancer drugs (*483–485*). Using β-isocupreidine (**568**) as a catalyst, the common intermediate **578** could be obtained in enantiopure form by reacting acrylate **569** with the O-protected aldehyde **577** (Scheme 120). Compound **578** could then be used to access the advanced intermediate **579**, which was then either converted into (+)-fostriecin (**575**) or (+)-phoslactomycin B (**576**) in a more step-linear sequence (*482*).

Desymmetrization of *meso*-compounds is a challenging but rewarding task. *Deng et al.* have established a method to desymmetrize the *meso*-anhydride **580** in the presence of catalytic amounts of the modified *Cinchona* alkaloid catalyst **581** to yield the hemiester **582** quantitatively and with very high enantiopurity (*486*). Further transformation into lactone **583** represents a short and efficient formal synthesis of (+)-biotin (**584**) (Scheme 121).

An organocatalytic desymmetrization of a *meso*-anhydride was also the starting point in a recent efficient protecting group-free total synthesis of the monoterpenoid

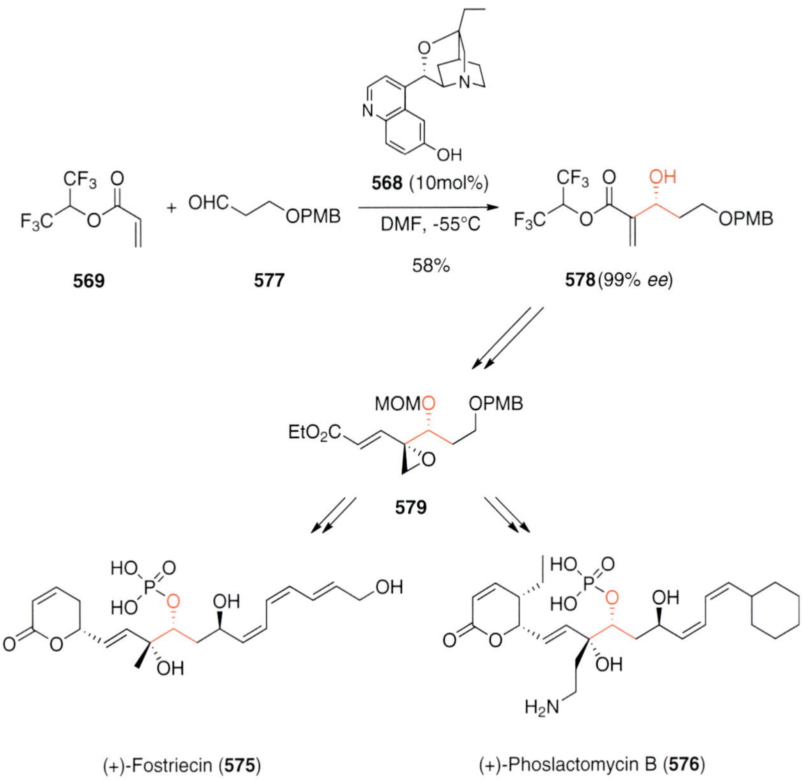

Scheme 120 Asymmetric *Morita-Baylis-Hillman* reaction used early in the syntheses of (+)-fostriecin (**575**) and (+)-phoslactomycin B (**576**)

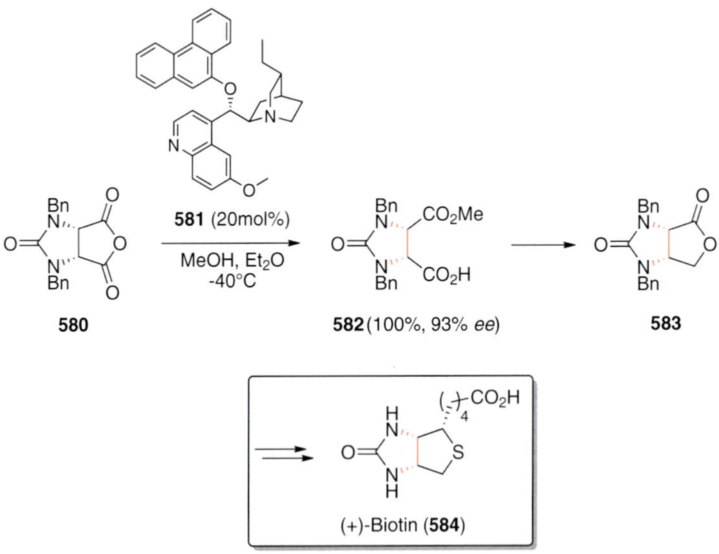

Scheme 121 Desymmetrization of *meso*-anhydrides using *Cinchona* alkaloid catalysts

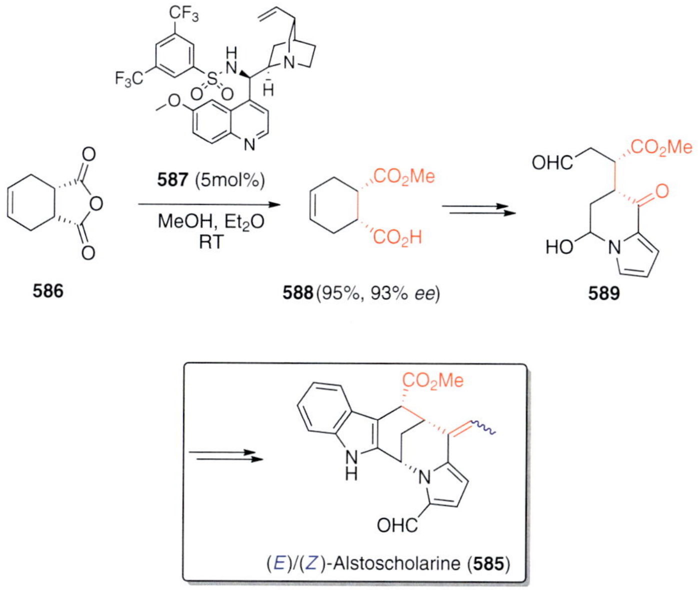

Scheme 122 Desymmetrization of *meso*-anhydrides in the protecting group-free total syntheses of (*E*)- and (*Z*)-alstoscholarine (**585**)

indole alkaloids (*E*)- and (*Z*)-alstoscholarine (**585**), by the group of *Zhu* (*487*). The complex alkaloids **585** were isolated from the leaves of *Alstonia scholaris*, a plant used in traditional medicine in South and Southeast Asia (*488*). The total synthesis commenced with a desymmetrization of *meso*-anhydride **586** using the bifunctional catalyst **587**. The hemiester **588** could thus be obtained in near-perfect yield (95%) and high enantioselectivity (93% *ee*) (Scheme 122) and served as the starting material of choice to obtain the advanced intermediate **589** in three more steps (*487*).

Over the years, *Cinchona* alkaloids have also been found to be efficient catalysts for enantioselective 1,4-addition reactions (*489*, *490*). Already by 1998, the group of *Terashima* had tested the *Cinchona* alkaloid-mediated 1,4-addition using dicarbonyl nucleophiles in their synthesis of the potent reversible acetylcholinesterase inhibitor huperzine A (**590**) (*489*). Despite being only modest in terms of enantioselectivity (<64% *ee*) and yield when using a stoichiometric amount of catalyst, this report indicated the potential of this methodology. As so often in the field of *Cinchona* alkaloid-based organocatalysis, it was left to the group of *Li Deng* to establish a highly selective and catalytic protocol for this type of transformation. In 2006, they reported a straightforward and highly selective synthesis of (+)-tanikolide (**591**) using an early-stage *Cinchona* alkaloid-catalyzed *Michael* addition to install the stereogenic center (*490*). Tanikolide (**591**) is an antifungal lactone from the marine cyanobacterium *Lyngbya majuscula* (*491*). After some screening, the catalyst **592** was identified as the best suited one for the addition of β-ketoester **593** to acrolein (**285**) to furnish aldehyde **594** in excellent enantiopurity (Scheme 123). Crude **594** was transferred directly further to yield (+)-tanikolide

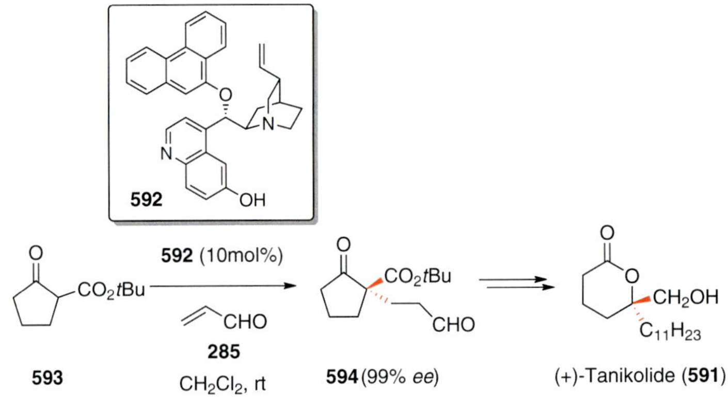

Scheme 123 Asymmetric *Michael* addition in the synthesis of (+)-tanikolide (**591**)

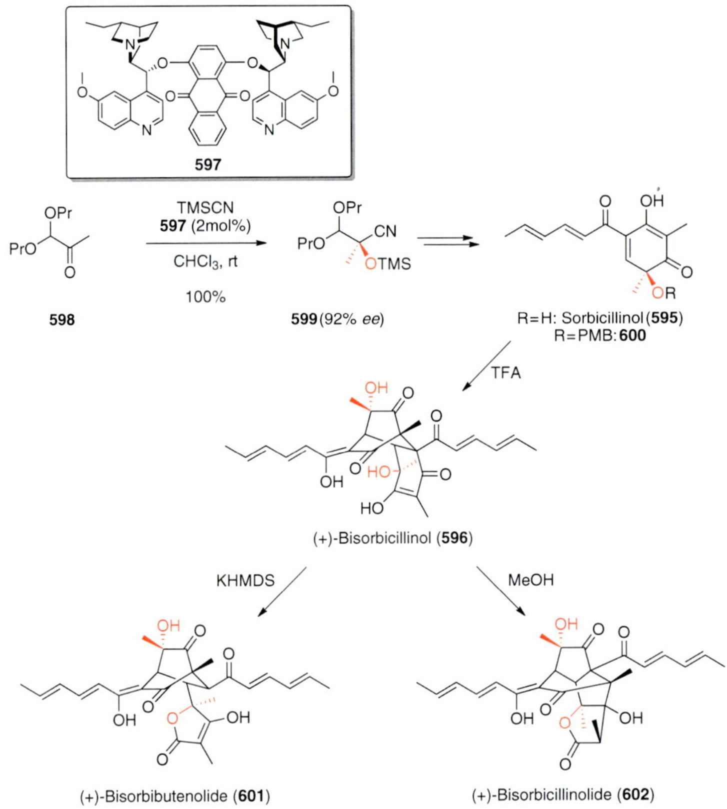

Scheme 124 Asymmetric cyanosilylation of ketones early in the syntheses of different members of the bisorbicillinoid family

(**591**) in just a few additional steps including a *Takai*-type olefination, oxidation state manipulations, and a *Baeyer–Villiger* oxidation (*490*).

Asymmetric additions of cyanides to carbonyl compounds represent efficient and versatile approaches for stereoselective carbon-carbon bond formations. The group of *Li Deng* has developed (Scheme 124) a high yielding and selective protocol for the asymmetric cyanosilylation of ketones by using dimeric *Cinchona* alkaloid catalysts (*492*). The bisorbicillinoids are a family of natural products isolated from various species of fungi with diverse biological properties (*493*). Although structurally diverse, they are proposed to originate from *Diels-Alder* dimerization of the common monocyclic precursor sorbicillinol (**595**) (*492*, *493*). Accordingly, *Diels-Alder* dimerization of **595** results in the formation of its dimer (+)-bisorbicillinol (**596**), which is believed to be also the key intermediate in the biosynthesis of other members of the family. Although structurally complex, the monomeric precursor **595** only contains one stereogenic center, which can be introduced by a *Cinchona* alkaloid **597**-catalyzed cyanosilylation of the starting material **598**. The corresponding cyanohydrin **599** was employed successfully to synthesize the protected sorbicillinol **600**. Acidification of **600** resulted in removal of the PMB-protecting group followed by immediate [4 + 2] dimerization to give (+)-bisorbicillinol (**596**). Interconversion of **596** into other members of the family is rather elegant and can be achieved either by treatment of **596** with one equivalent of base to give bisorbibutenolide (**601**), or by prolonged treatment with methanol to furnish the rearranged bisorbicillinolide (**602**) (*492*, *493*).

8.2 Phosphine Catalysis

Phosphines are another class of nucleophilic catalysts that have attracted considerable interest over the last several years (*494–497*). Interestingly, although a number of impressive asymmetric reactions using well-designed chiral phosphine catalysts have been reported in the past (*498–500*), applications in natural product syntheses have so far been carried out mainly employing achiral phosphines, either giving racemic products (*501–503*), or performing the reaction in a stereospecific fashion with an appropriate chiral starting material (*504–506*). Although it is not totally within the scope of this volume to cover applications of achiral organocatalysts, three very impressive and illustrative examples using achiral phosphine catalysts in a stereospecific fashion in complex total syntheses are described briefly. It is felt that these give a good overview concerning the substantial potential of this methodology in general and, in addition, it seems reasonable to propose that the use of chiral phosphines in natural product synthesis will receive increased attention in the future.

In 2003, *Lu* and *Du* reported the first total synthesis of (−)-hinesol (**603**) (*504*). Hinesol is an important component of the Chinese drugs *Chang Zhu* (*Atractylodes lancea* var. *chinensis*) and *Baizhu* (*Alpinia japonica*) and shows spasmolytic and

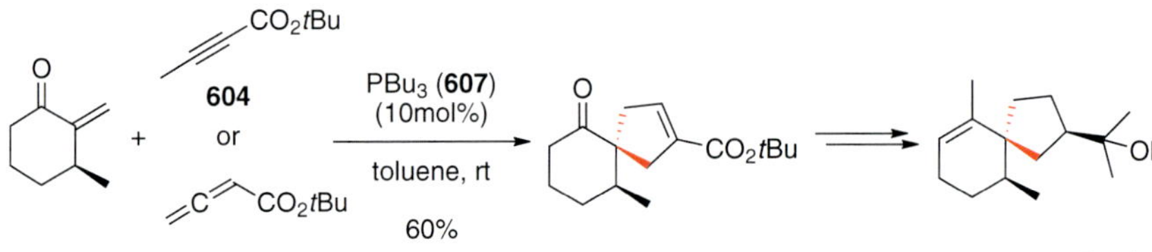

Scheme 125 Phosphine-catalyzed [3 + 2] cycloaddition in the total synthesis of (−)-hinesol (**603**)

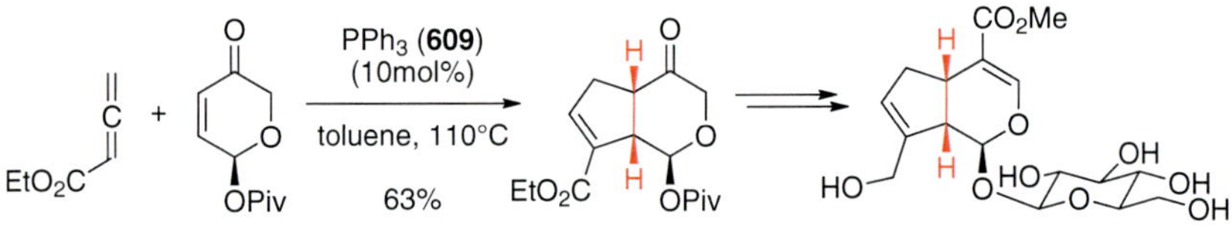

Scheme 126 Phosphine-catalyzed [3 + 2] cycloaddition in the total synthesis of (+)-geniposide (**612**)

antigastric ulcer activity as well as specific inhibition of H^+- and K^+-ATPase (*507, 508*). Synthesis of this interesting target in a stereoselective fashion could be achieved by carrying out a [3 + 2] cycloaddition of either alkyne **604** or allene **605** with chiral enone **606** in the presence of tributylphosphine (**607**) to furnish the spirocyclic compound **608** in reasonable yield (60%) and excellent enantiopurity (94% *ee*) (Scheme 125).

A triphenylphosphine (**609**)-catalyzed [3 + 2] cycloaddition between allene **610** and chiral enone **611** was also the key step in a recent synthesis of the iridoid β-glucoside (+)-geniposide (**612**) by the *Krische* group (*506*). Isolated from *Gardenia jasminoides* (*509*), (+)-geniposide (**612**) displays potential antitumor and anti-inflammatory activities (*510, 511*). Carrying out the cycloaddition at elevated temperature (110°C) gave the bicyclic product **613** in 63% yield as a single regio- and stereoisomer, thus paving the way towards (+)-geniposide (**612**) (Scheme 126) (*506*).

The spinosyns are a family of polyketide natural products generated by *Saccharopolyspora spinosa* possessing potent insecticidal activity (*512*). Some years ago the group of *Roush* accomplished the synthesis of its most prominent member (−)-spinosyn A (**614**) (*505*). After several preliminary steps, the tricyclic intermediate **615** could be obtained from the easily available starting materials **616** and **617**. Carrying out an intramolecular vinylogous *Morita-Baylis-Hillman* reaction catalyzed by trimethylphosphine (**618**) gave the targeted cyclization product **619** accompanied with small amounts of double bond migration and epimerization by-products (<12% in total) (Scheme 127). After additional transformations, (−)-spinosyn A (**614**) could be obtained in reasonable yield and was identical in all spectroscopic properties on comparison with a sample isolated from natural

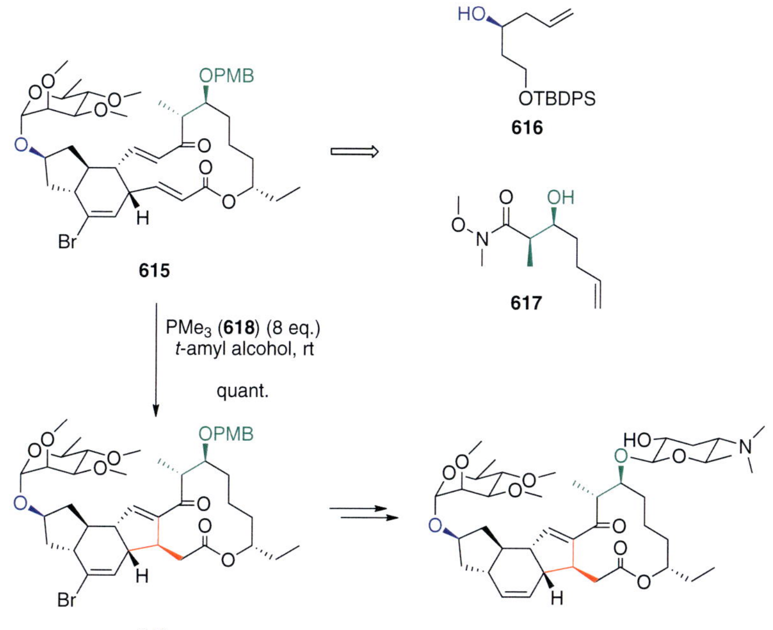

Scheme 127 Phosphine-catalyzed *Morita-Baylis-Hillman* reaction in the total synthesis of (–)-spinosyn A (**614**)

sources. This is a natural product synthesis where the organocatalytic step is just one interesting reaction in a whole series of highly impressive transformations necessary to produce the target compound so the original report by *Roush et al.* is recommended to the interested reader (*505*).

8.3 Carbene Catalysis

The use of carbenes as asymmetric organocatalysts has attracted the interest of more and more research groups over the last few years (*513–515*). While for a long time the application of carbenes in asymmetric natural product syntheses has been limited mainly to their (very important) role as chiral ligands in metal-catalyzed transformations (*e.g.* metathesis (*516*)), the use of (chiral) carbenes as nucleophilic catalysts has only found its place slowly in (asymmetric) natural product syntheses (*517–521*). In principle, the situation is similar to that of phosphine catalysis (see the previous section), with achiral catalysts being used routinely for different reactions including natural product syntheses, whereas chiral carbene catalysts are still something of a curiosity within this context. (Several examples describing the use of achiral carbenes in natural product synthesis date back to the 1970s (*522*,

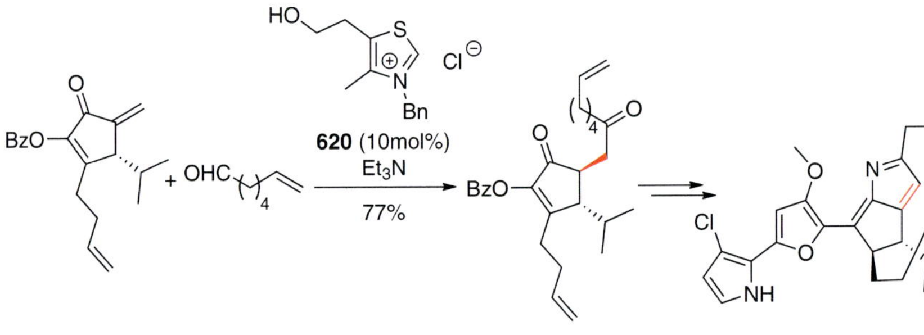

621 622 623 Roseophilin (**624**)

Scheme 128 Achiral NHC-catalyzed *Stetter* reaction in the synthesis of roseophilin (**624**)

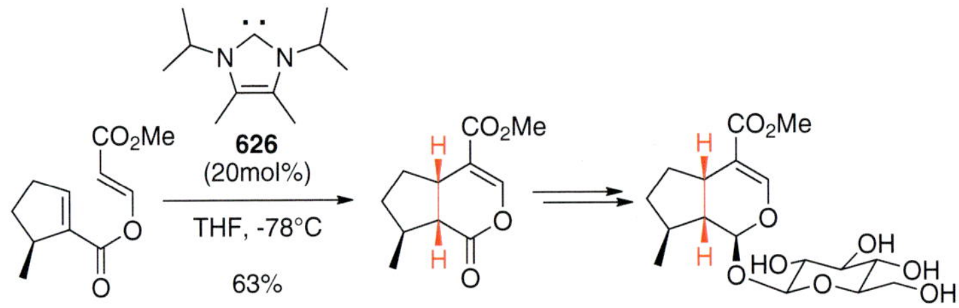

627 628 7-Deoxyloganin (**625**)

Scheme 129 Synthesis of 7-deoxyloganin (**625**)

523)). To illustrate the substantial potential of this methodology, which should play an increasingly important role in the future, two very interesting examples for the application of achiral N-heterocyclic carbene (NHC)-catalyzed reactions on chiral substrates in natural product synthesis will be discussed.

In 2001, *Tius et al.* used thiazolium chloride **620** as a carbene precursor in a diastereoselective intermolecular *Stetter* reaction (*518*) between the chiral donor **621** and aldehyde **622**. The 1,4-diketone **623** thus obtained was then used successfully to access roseophilin (**624**) (Scheme 128) (*517*). The promising cytotoxicity of roseophilin (**624**), a macrocyclic pigment isolated from *Streptomyces griseoviridis* (*524*), has attracted considerable attention resulting in a number of partial and total syntheses of this alkaloid (*525*), and this example demonstrates nicely how (even achiral) organocatalysts can play an important role in obtaining a complex natural product.

The potential importance of the iridoid family of compounds was already mentioned previously in this volume (see Scheme 48 and 126). The *Lupton* group has reported recently the total synthesis of the natural product 7-deoxyloganin (**625**) (*521*). Synthesis of the bicyclic core was achieved by a NHC (**626**)-catalyzed rearrangement of the precursor **627** to furnish the lactone **628** in reasonable yield and with full transfer of stereo-information (Scheme 129), thus giving access to **625** after just three further steps.

8.4 Synopsis

Chiral base catalysis is one of the most versatile and broadly applicable types of catalysis. In particular, the potential of tertiary amines to act both as a base and as a nucleophilic catalyst makes chiral tertiary amines like *Cinchona* alkaloids a privileged catalyst structure in modern synthesis chemistry. In addition, the field of achiral phosphine and carbene catalysis has proven its potential in numerous applications in the past and it is probably only a matter of time until chiral phosphines and carbenes will also be used routinely for other presently demanding natural product total synthesis (Table 7).

Table 7 (A)chiral bases and nucleophilic catalysts used in asymmetric natural product syntheses

Catalyst	Product	References
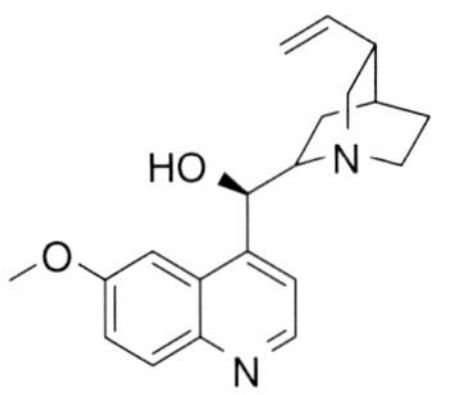 **2**	(+)-Calanolide A (**535**) (+)-Inophyllum B (**536**) UCS1025A (**543**)	(*451*) (*451*) (*454*)
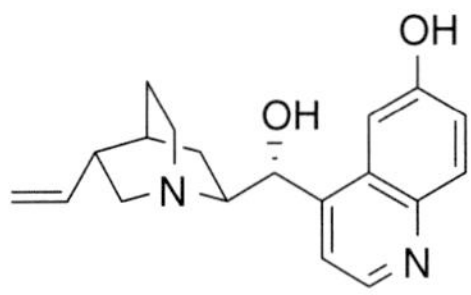 **546**	(−)-Spirobrassinin (**544**)	(*458*)
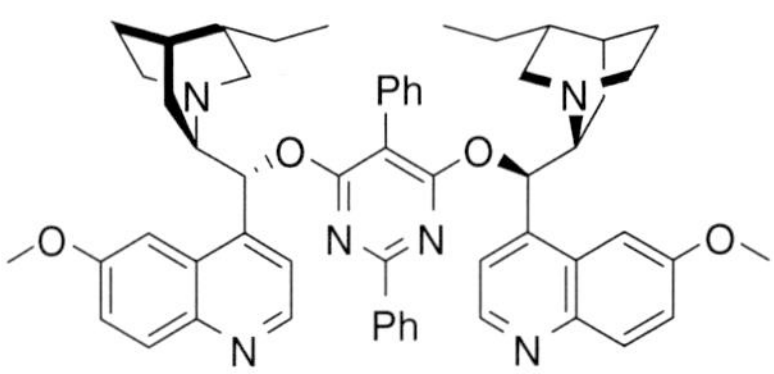 **561**	Eicosanoid (**555**) Solandelactone A and B intermediate **556**	(*473*) (*474*)
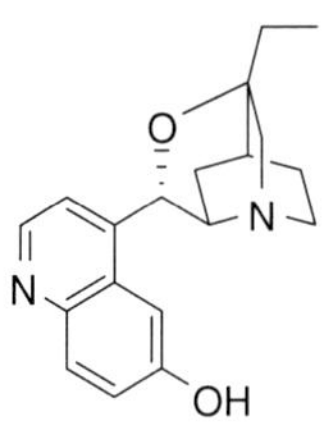 **568**	Epopromycin B (**573**) (+)-Fostriecin (**575**) (+)-Phoslactomycin B (**576**)	(*479*) (*482*) (*482*)
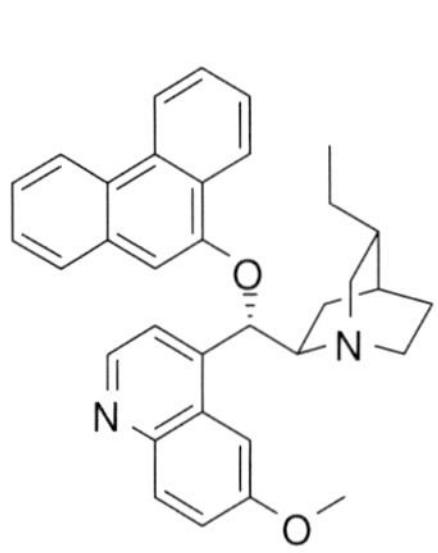 **581**	(+)-Biotin (**584**)	(*486*)

(continued)

Table 7 (continued)

Catalyst	Product	References
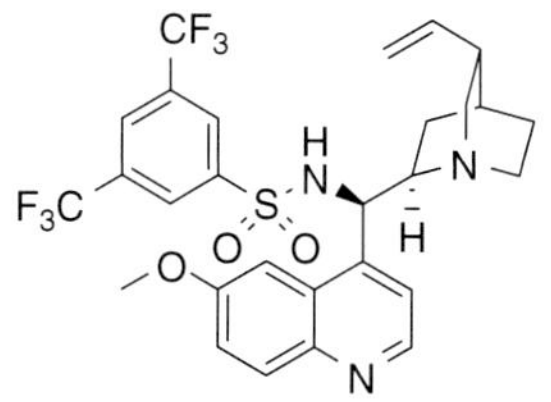 **587**	Alstoscholarine (**585**)	(*487*)
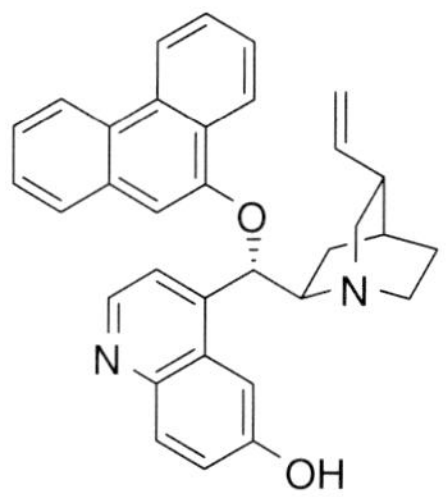 **592**	(+)-Tanikolide (**591**)	(*490*)
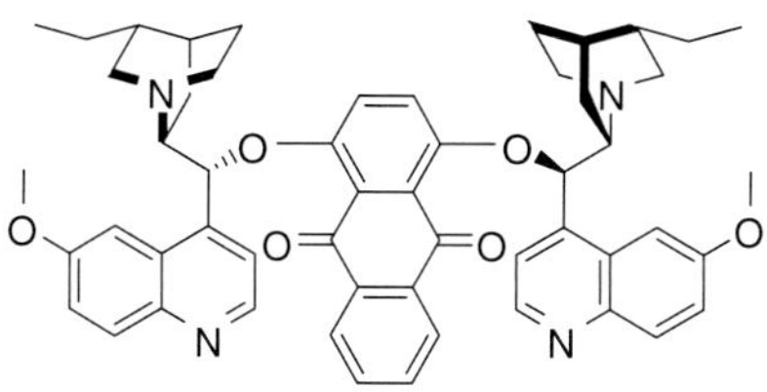 **597**	(+)-Bisorbicillinol (**596**) Bisorbibutenolide (**601**) Bisorbicillinolide (**602**)	(*492*) (*492*) (*492*)
PBu_3 (**607**) PPh_3 (**609**) PMe_3 (**618**)	(−)-Hinesol (**603**) (+)-Geniposide (**612**) (−)-Spinosyn A (**614**)	(*504*) (*506*) (*505*)
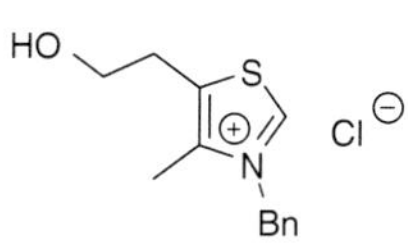 **620**	Roseophilin (**624**)	(*517*)
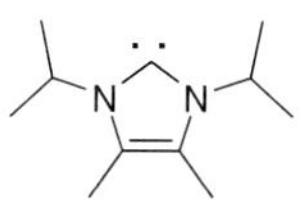 **626**	7-Deoxyloganin (**625**)	(*521*)

9 Asymmetric Oxidations and Reductions

The importance of asymmetric oxidations and reductions was clearly demonstrated in 2001 when the *Nobel* prize in chemistry was awarded to *K. Barry Sharpless* (*526*), *William S. Knowles* (*527*), and *Ryoji Noyori* (*528*) for their groundbreaking contributions to this field.

Historically, both oxidations and reductions have been dominated by transition metal catalysis. However, besides the use of metal catalysts, a variety of metal-free oxidation (*529*) and reduction processes have been developed. Some examples concerning the successful application of chiral secondary amine catalysts in oxidation reactions *via* enamine catalysis (*111–113*, *116*, *120*) as well as *via* iminium catalysis (*292*, *293*, *296*) have been discussed in earlier chapters (for enamine-catalyzed α-oxygenations see Sect. 2.3.1 and for iminium-catalyzed oxygenations see Sect. 3.2.5). In addition, the use of iminium catalysis to facilitate highly asymmetric transfer hydrogenations has been covered in a previous chapter (Sect. 3.2.1) (*203–205*, *210*). It was also mentioned earlier that asymmetric phase-transfer catalysis provides a useful tool for stereoselective epoxidation reactions (*333*). Thus, it is apparent that asymmetric organocatalytic oxidation and reduction reactions are rather diverse with respect to activation mode and the type of catalyst employed. It is not the aim of the present chapter to give a comprehensive overview concerning different organocatalytic redox state manipulation reactions but rather to highlight some of the most versatile methods by discussing several successful applications in natural product syntheses in greater detail.

9.1 Organocatalytic Oxidations

Besides the secondary amine-catalyzed oxygenation reactions (see Sects. 2.3.1. and 3.2.5.) and phase-transfer catalyzed epoxidations (Chap. 6.) already mentioned, asymmetric epoxidation reactions using the method developed by *Shi et al.* (*530*) have found to be highly useful in complex total syntheses (*531–535*). The *Shi* epoxidation employs the fructose-derived ketone **629** as an easily available natural

M. Waser, *Asymmetric Organocatalysis in Natural Product Syntheses*,
Progress in the Chemistry of Organic Natural Products 96,
DOI 10.1007/978-3-7091-1163-5_9,

chiral pool-based catalyst to achieve asymmetric epoxidations of a variety of functionalized and unfunctionalized alkenes on treatment with potassium peroxymonosulfate ($KHSO_5$, Oxone) under strict control of pH (*536–538*). In general, the oxidant causes formation of a chiral oxirane by reacting with the catalyst, which then causes a face-selective oxygen transfer to the olefin, thus regenerating the catalyst.

The high versatility of this protocol was demonstrated in an elegant total synthesis of the chiral C_2-symmetric pentacyclic oxasqualenoid glabrescol (**630**) by *E.J. Corey et al.* (*531*). Glabrescol is a pentaoxacyclic squalenoid, which has been isolated from the Caribbean plant *Spathelia glabrescens* and is a biosynthesis precursor for a number of possible steroids and triterpenoids (*539*). To access **630**, epoxidation of precursor **631** using the D-fructose-derived *Shi* catalyst **629** (1.5 eq.) afforded the tetraepoxide **632** in 66% yield with a diastereomeric purity of approximately 80% and an estimated (*R*):(*S*) selectivity at each double bond of >20:1 (Scheme 130) (*531*, *540*). Further steps then yielded glabrescol (**630**) and also resulted in a revision of the originally postulated structure.

Another example showing the versatility of this methodology was reported by *Morimoto et al.* in 2005 in their total synthesis of (+)-aurilol (**633**) (*535*). Aurilol is a bromotriterpene polyether that was isolated from the sea hare *Dolabella auricularia* by *Yamada et al.* in 1998, exhibiting cytotoxicity against HeLa S3 cells (IC_{50} = 4.3 μg/cm^3) (*541*). In their first total synthesis of this demanding target, *Morimoto et al.* employed a *Shi* epoxidation on olefin **634** using the

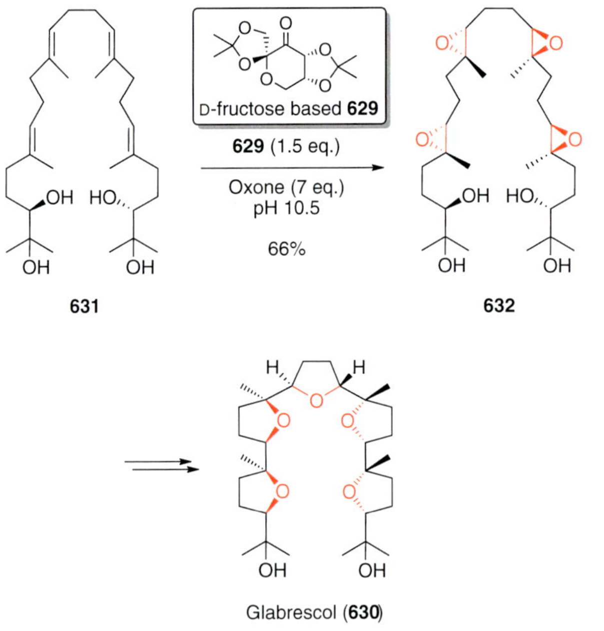

Scheme 130 Asymmetric *Shi* epoxidation in *Corey's* total synthesis of glabrescol (**630**)

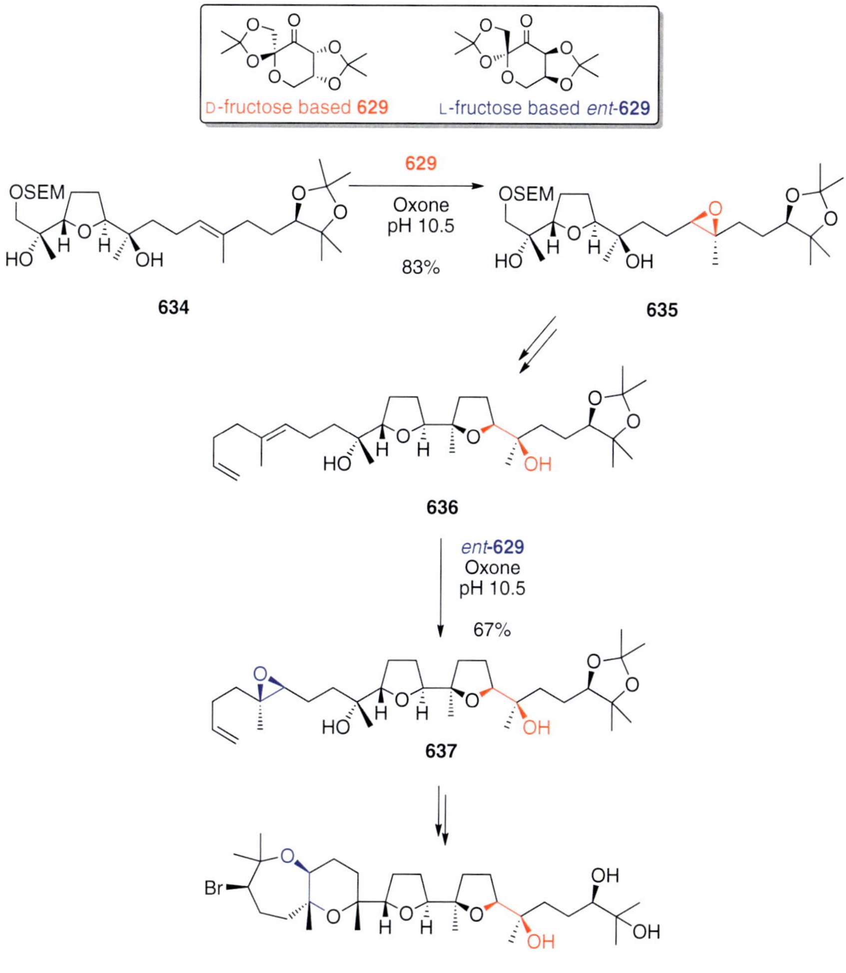

Scheme 131 Asymmetric *Shi* epoxidations in the total synthesis of (+)-aurilol (**633**)

D-fructose derived catalyst **629** to obtain initially intermediate **635** in 83% yield. Later in the sequence, the advanced intermediate **636** was epoxidized in the presence of the enantiomeric L-fructose-derived catalyst *ent*-**629** to furnish oxirane **637** (Scheme 131). After further transformations the first total synthesis and the complete assignment of the stereochemistry of this complex natural product was achieved.

Besides the *Shi* epoxidation, which has also proven its potential in a series of other complex syntheses (*532*, *534*, *542*), the *Julia-Colonna* asymmetric epoxidation (*543–545*) using polyamino acid catalysts also has been found to be a versatile method in total synthesis (*546–548*). In 1998, the group of *Roberts* reported a

Julia-Colonna epoxidation in the synthesis of (+)-clausenamide (**638**). The γ-lactam clausenamide (**638**) was isolated from the leaves of *Clausena lansium*, a species that is a liver-protecting Chinese folk medicine and is utilized also in cases of acute and chronic viral hepatitis (*549*). In their synthesis of **638**, *Roberts et al.* employed a fixed-bed poly-L-leucine catalyst together with DABCO-H_2O_2 as the oxidant to carry out an early-stage epoxidation of chalcone **639** to obtain epoxide **640** in high enantiomeric excess (>98% *ee*) (Scheme 132). After five more steps, the total synthesis of (+)-clausenamide (**638**) was accomplished in an overall yield of 40% (*547*).

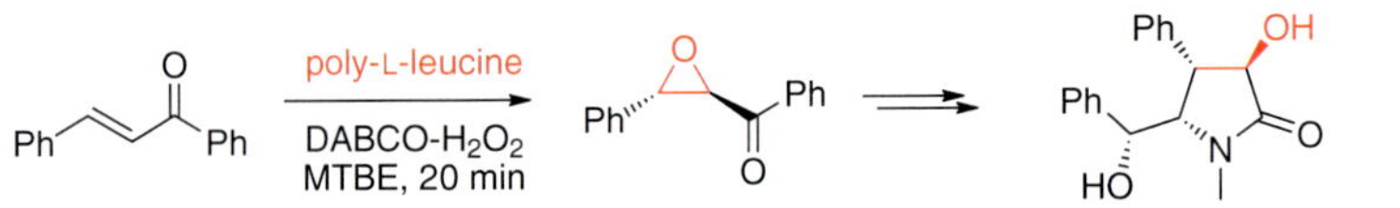

639 **640** (>98% *ee*) (+)-Clausenamide (**638**)

Scheme 132 Asymmetric *Julia-Colonna* epoxidation in the synthesis of (+)-clausenamide (**638**)

The same group also successfully applied a *Julia-Colonna* epoxidation to the syntheses of the naturally occurring styryl lactones (+)-goniofufurone (**641**), (+)-goniopypyrone (**642**), and (+)-8-acetylgoniotriol (**643**) (*548*). These compounds were isolated from the Malaysian *Goniothalamus giganteus* and possess cytotoxic activity against human tumor cells (*550*, *551*). On carrying out the asymmetric epoxidation of furyl styryl ketone **644** with urea-H_2O_2 as the oxidant and poly-L-leucine as the catalyst, the key intermediate **645** could be obtained quantitatively and with high enantiopurity (Scheme 133). Epoxide **645** served well to furnish the natural products (+)-goniofufurone (**641**), (+)-goniopypyrone (**642**), and (+)-8-acetylgoniotriol (**643**) after a few additional steps (*548*).

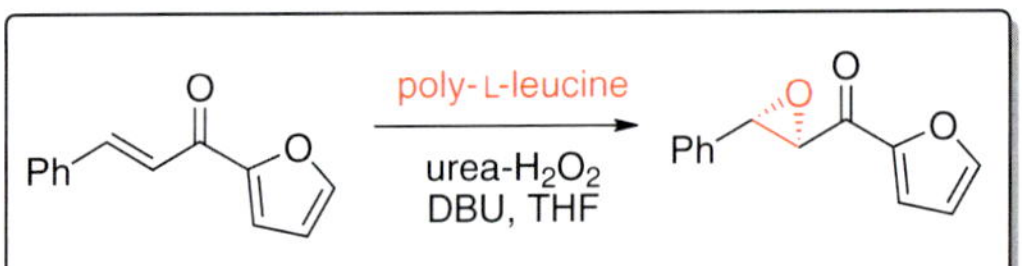

644 >95% **645** (>99% *ee*)

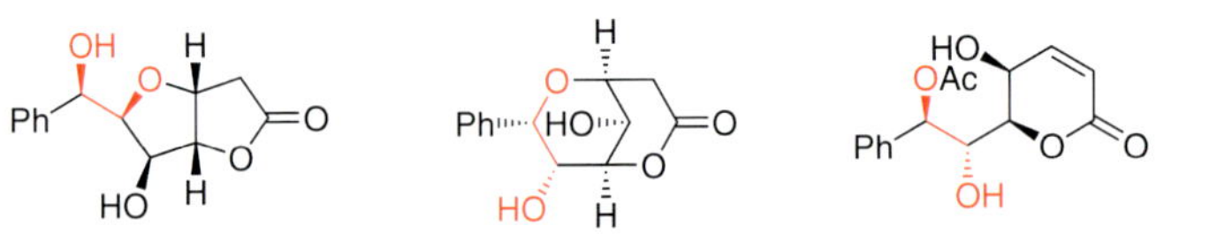

(+)-Goniofufurone (**641**) (+)-Goniopypyrone (**642**) (+)-8-Acetylgoniotriol (**643**)

Scheme 133 Asymmetric *Julia-Colonna* epoxidation in the syntheses of (+)-goniofufurone (**641**), (+)-goniopypyrone (**642**), and (+)-8-acetylgoniotriol (**643**)

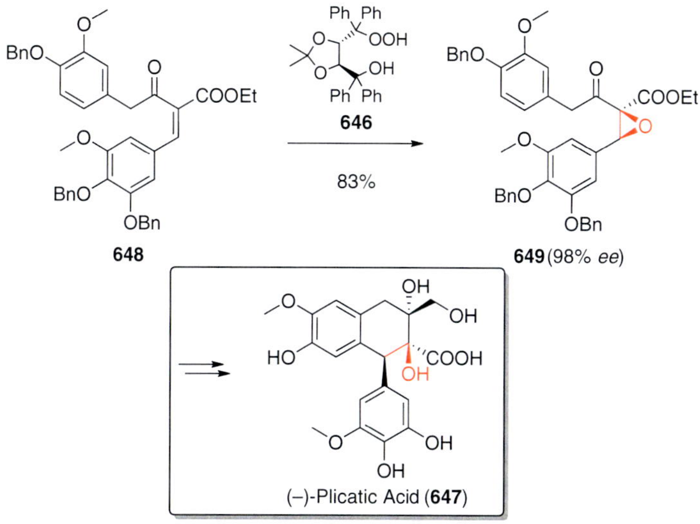

Scheme 134 Asymmetric epoxidation in the synthesis of (−)-plicatic acid (**647**)

Besides employing small-organic molecule catalysts to facilitate oxidations using achiral oxidants, the use of chiral oxidants like tartaric acid-derived **646** has provided a useful method for several applications (*552–554*). Although not a catalytic procedure, one example may be mentioned since it highlights impressively the potential of this methodology. The group of *Deng* reported recently a relatively short (12 step) and high-yielding (14% overall yield) total synthesis of (−)-plicatic acid (**647**) (*554*). Plicatic acid was isolated in 1959 from western red cedar *Thuja plicata* (*555*) and has been found to be a causative agent of occupational asthma as well as resulting in inflammatory and allergic reactions (*556–558*). *Deng's* retrosynthetic analysis required a stereoselective epoxidation of the trisubstituted electron-deficient olefins **648**. In general, nucleophilic epoxidation of such a target is known to be a rather challenging transformation (*554*). After careful investigation it was discovered that the use of *Seebach's* tartaric acid-derived hydroperoxide **646** (*552*, *553*) allowed them to obtain the targeted epoxide **649** in high yield and with excellent enantiomeric excess (Scheme 134). Further transformations then resulted in the first total synthesis of this long-known target in an impressively high overall yield of 14% (*554*).

Recently *Iwabuchi et al.* reported the first total synthesis of (−)-idesolide (**650**) (*559*), an efficient inhibitor of nitric oxide production induced by lipopolysaccharide in BV2 microglial cells, isolated from the deciduous tree *Idesia polycarpa* (*560*). As idesolide may originate from dimerization of the precursor **651**, two major challenges had to be addressed in this investigation. The first was installation of the required absolute configuration in **651** and the second, identifying useful dimerization conditions for this monomeric precursor. Synthesis of **651** could be achieved successfully by carrying out an oxidative kinetic resolution on the easily synthesized racemic

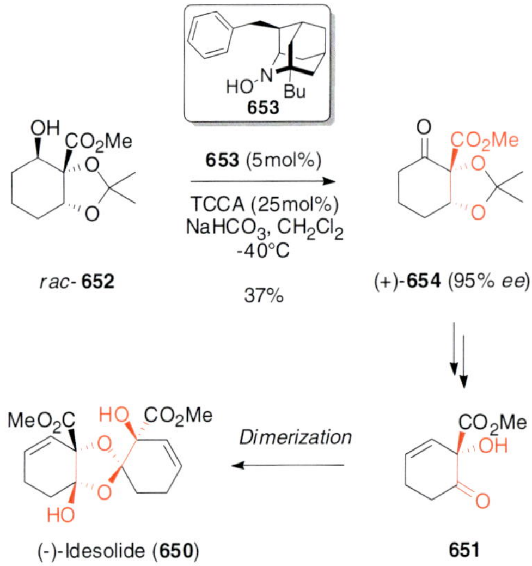

Scheme 135 Organocatalytic oxidative kinetic resolution in the synthesis of (−)-idesolide (**650**)

sec-alcohol **652** using catalytic amounts of the chirally modified 2-azaadamantane *N*-oxyl **653** (*561*) together with trichloroisocyanuric acid (TCCA) (Scheme 135). Followed by further transformations, the dimerization precursor **651** was obtained successfully in high enantiopurity. Final dimerization then resulted in the first total synthesis of the spiro compound (−)-idesolide (**650**) (*559*).

As already mentioned at the beginning of this chapter, it is not intended to give an in-depth coverage of asymmetric oxidation reactions using either organocatalysts or, as shown in the second-to-last example, stoichiometric chiral organic oxidants, but only to highlight the potential of these methods by presenting some selected examples. One criticism that might be raised is that these methodologies sometimes require stoichiometric quantities of the catalyst (see Scheme 130). Nevertheless, the easy availability of these powerful oxidation catalysts and the fact that reactions sometimes can be carried out when otherwise barely possible (see Scheme 134), makes them useful alternatives to conventional (transition-) metal-catalyzed processes.

9.2 Organocatalytic Reductions

The use of iminium-catalysis to facilitate highly asymmetric transfer hydrogenations has already been covered in some detail previously (Sect. 3.2.1) (*203–205*, *210*) and the interested reader is referred to the original literature cited therein. Besides iminium activation in combination with *Hantzsch* dihydropyridines as hydride donors, the use of chiral phosphoric acids in combination with *Hantzsch* ester-based hydride donors (Sect. 7.1) has resulted in some useful applications over the last few years (*376*, *379*).

In addition to these two activation modes, the field of organocatalytic reductions is more a question of definition, since one might argue if applications using boranes or silanes are organocatalytic or not. It is not the intention of the author to define this matter, since in the last few years it has been demonstrated that one can not differentiate clearly between the different classes of catalysis (*i.e.*, metal catalysis *versus* organocatalysis *versus* enzymes).

Based on the fact that both boranes and silanes belong to the most versatile reducing agents available to synthesis-oriented organic chemists, considerable effort in the field of asymmetric reductions using these hydride-donors has been undertaken. For example, in the past, silanes have been investigated as easily available achiral reducing agents in combination with either chiral *Lewis* bases (*562*) or chiral phase-transfer catalysts (*563*). While phase-transfer catalyzed approaches have so far been only modestly successful in terms of enantioselectivity (<70% *ee*), chiral formamides have been reported to give reasonable enantiomeric excesses (>90% *ee*) in a variety of test reactions (*563*).

The use of boranes has attracted considerably more interest over the last several decades and particularly the use of chiral oxazaborolidine catalysts in combination with borane as a stoichiometric reducing agent (the well-known *Corey-Bakshi-Shibata* (*CBS*) reduction) (*564*, *565*) represents one of the most important and versatile methods for the stereoselective reduction of prochiral ketones. It is not intended to give a detailed overview of the successful applications of this versatile methodology for complex (natural product) synthesis, as this would be far beyond the scope of this volume. Instead two examples are chosen below to highlight the potential of this method, especially when used for highly selective late-stage modifications (*566*, *567*).

The first of these examples was reported (Scheme 136) by the group of *E. J. Corey* (*566*). In 1997, this group published the total synthesis of the *neo*-isolabdanoid sesterterpene dysidiolide (**655**) from the Caribbean sponge *Dysidea etheria* (*568*). Dysidiolide was found to be the first naturally derived inhibitor of

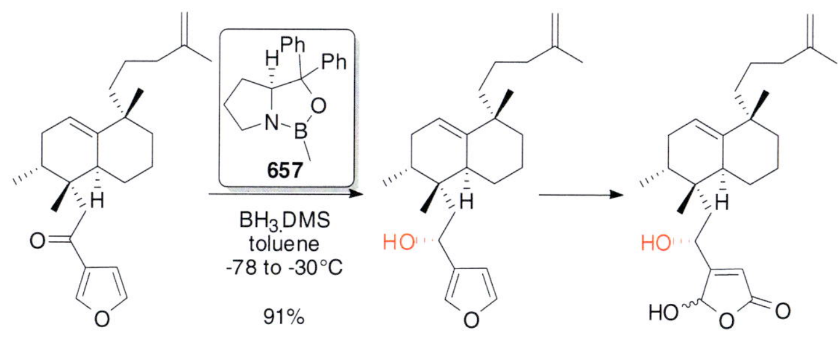

656 **658** (-)-Dysidiolide (**655**)

Scheme 136 *CBS*-reduction in the synthesis of dysidiolide (**655**)

human cdc25A protein phosphatase ($IC_{50} = 9.4\ \mu M$) (*568*). In their total synthesis of **655**, *Corey et al.* employed a late-stage reduction of ketone **656** in the presence of the (*S*)-configured catalyst **657** using the dimethylsulfide complex of BH_3 as the stoichiometric reducing agent to give the advanced intermediate **658** in high yield and with excellent diastereoselectivity, leaving only one step to accomplish the total synthesis of dysidiolide (**655**) (*566*).

A second late-stage application of the *CBS*-reduction was reported by *Forsyth et al.* in their total synthesis of okadaic acid (**659**) (*567*). This cytotoxic polyether was isolated initially from the marine sponge *Halichondria okadai* in 1981 (*569*). In their total synthesis of this challenging target, *Forsyth et al.* carried out a *CBS*-reduction on the highly functionalized enone **660**. The crude reduction product was directly subjected to an acid-catalyzed spiroketalization to furnish the advanced intermediate **661** in 81% yield (over both steps) (Scheme 137). Final protecting group manipulations were used to accomplish the total synthesis of okadaic acid (**659**) (*567*).

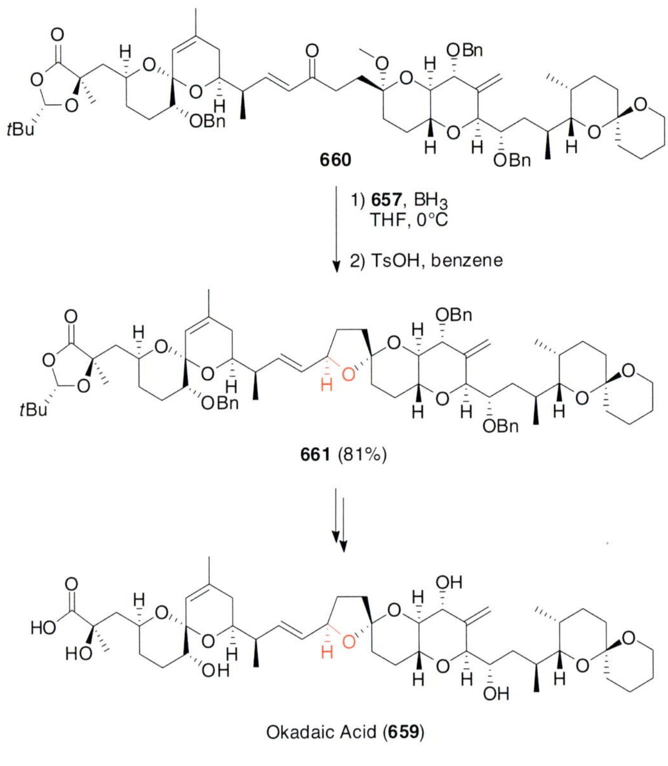

Scheme 137 *CBS*-reduction in the total synthesis of okadaic acid (**659**)

9.3 Synopsis

The field of redox catalysis is the most widespread among the different fields covered in this contribution. The great importance of stereoselective redox state manipulation reactions makes these types of transformations extraordinarily important for organic chemists. Over the last few decades the use of organic molecule catalysts to carry out stereoselective oxidations and reductions has received increased attention and has found a well-deserved place along with metal-catalyzed approaches.

It was mentioned at the beginning of this section that only selected examples would be presented and some successful organocatalytic oxidation or reduction steps in natural product syntheses were covered in the earlier chapters of this volume. To provide as much comprehensive information as possible, Tables 8 and 9 are intended to give summaries of the examples covered in this section and also of those discussed in the earlier sections. Of course, not only hydrogenation or oxygen-introducing reactions but also amination reactions or some of the other transformations covered in this volume are redox-type reactions. However, for the purposes of clarity, only oxygen-introducing reactions as well as oxidations of hydroxy groups are included in Table 8.

Table 8 Selected examples of organocatalytic oxidations in asymmetric natural product syntheses

Catalyst	Product	References
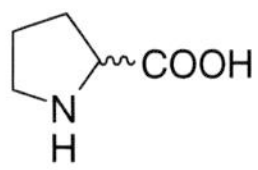 **12**	RK-805 (**129**)	(*112*)
	(+)-Panepophenanthrin (**131**)	(*113*)
	Neosymbioimine (**134**)	(*116*)
	Disparlure (**135**)	(*120*)
	(−)-Brasoside (**198**)	(*115*)
	(−)-Littoralisone (**199**)	(*115*)
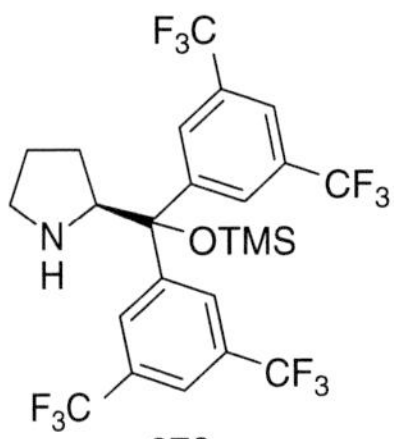 **276**	Pheromone **319**	(*297*)
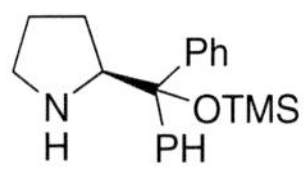 **138**	Hirsutellone B (**313**)	(*293*)
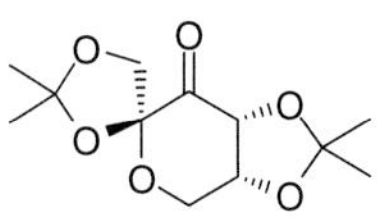 **629**	Glabrescol (**630**)	(*531*)
	(+)-Aurilol (**633**)	(*535*)
poly-L-leucine	(+)-Clausenamide (**638**)	(*547*)
	(+)-Goniofufurone (**641**)	(*548*)
	(+)-Goniopypyrone (**642**)	(*548*)
	(+)-8-Acetylgoniotriol (**643**)	(*548*)
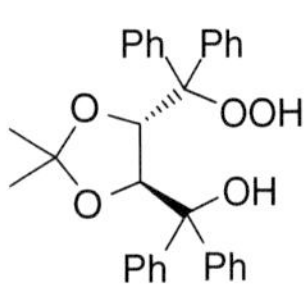 **646**	(−)-Plicatic acid (**647**)	(*554*)

(continued)

Table 8 (continued)

Catalyst	Product	References
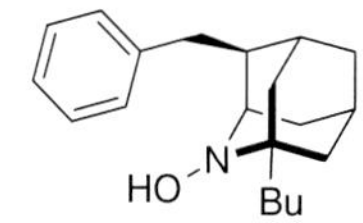 **653**	(−)-Idesolide (**650**)	(*559*)

Table 9 Selected examples of organocatalytic reductions in asymmetric natural product syntheses

Catalyst	Product	References
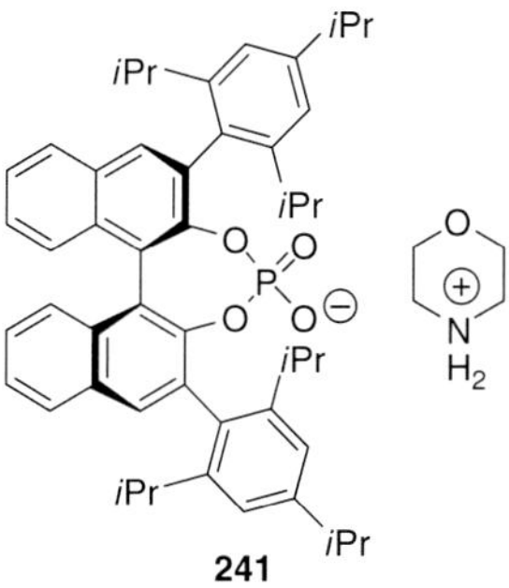 **241**	Pheromone (*R*)-**244**	(*210*)
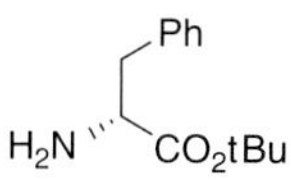 **246**	(−)-Platensimycin (**245**)	(*213*)
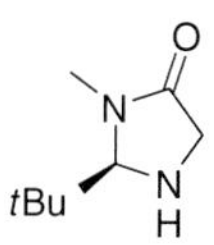 **253**	Dysideaproline E (**256**)	(*221*)
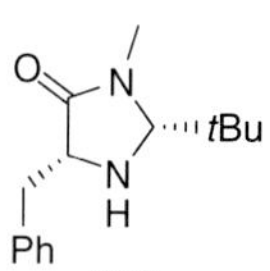 **228**	(+)-Ricciocarpin A (**336**)	(*303*)

(continued)

Table 9 (continued)

Catalyst	Product	References
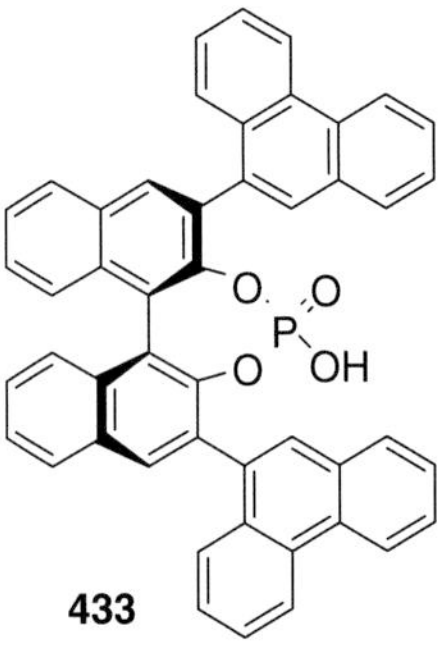 **433**	(+)-Galipinine (**436**) (+)-Cuspareine (**437**) (−)-Angustureine (**308**)	*(376)* *(376)* *(376)*
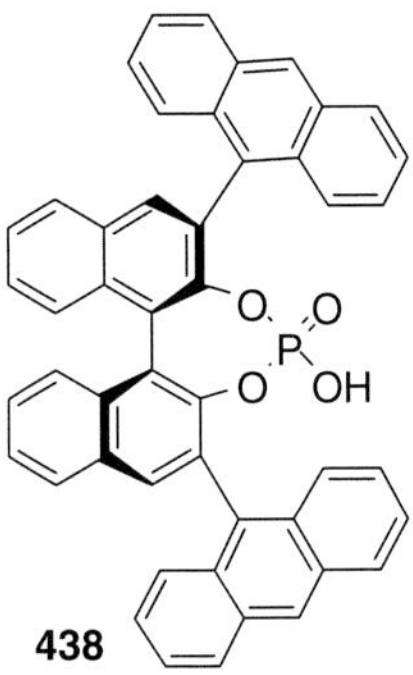 **438**	*diepi*-Pumiliotoxin C (**441**)	*(379)*
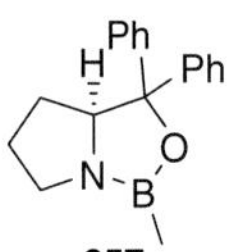 **657**	Dysidiolide (**655**) Okadaic acid (**659**)	*(566)* *(567)*

10 Conclusions

The challenging complexities of biologically active natural products synthesis require powerful and efficient methods to gain access to such compounds in an efficient and (if possible) highly stereoselective way. The field of asymmetric (organo)-catalysis has made dramatic progress over the last decade, as can be seen not only from the examples that have been presented in this contribution to the series, but also in looking at the large numbers of new catalysts, organic transformations, and natural product syntheses reported on almost a daily basis. It was mentioned earlier in this volume that it does not seem useful to differentiate strictly between different types of catalysis (*e.g.* metal catalysis *versus* organocatalysis), since almost all of the successful natural product syntheses shown in this contribution required additional, very often metal-catalyzed, transformations to achieve the final goal. In addition, and perhaps most importantly, the majority of synthesis-oriented chemists may not be concerned whether a method is believed to be organocatalytic or of some other type, so long as it fulfills its purpose in an efficient and reliable way. Moreover, it has been shown in some examples that the well-balanced interplay between organic molecule catalysts and metal catalysts enables the construction of complex molecular architectures in a strikingly impressive and efficient manner. Thus, in the opinion of the author there will be an increasing demand for the development of even more combined cascade-type approaches employing different catalysts, and different catalytic modes to provide the chemical community new tools to access reliably such highly interesting natural products.

With respect to the particular methods described in this contribution, one can state with confidence that enamine and iminium catalysis have already gathered a well-deserved place in the synthetic chemist's armamentarium, as has been illustrated by several successful applications. These catalytic methodologies have been found to benefit from simple operational procedures, high enantioselectivities, and broad substrate scope. The same can be said for organocatalytic redox reactions, chiral bases, and chiral hydrogen bonding donors, which have all had substantial use for several decades. Surprisingly, other interesting methods like asymmetric phase-transfer catalysis and SOMO catalysis have thus far been less

M. Waser, *Asymmetric Organocatalysis in Natural Product Syntheses*,
Progress in the Chemistry of Organic Natural Products 96,
DOI 10.1007/978-3-7091-1163-5_10,

widely used for complex natural product synthesis. However, it seems reasonable that these will become more prominent for future applications.

Finally, one of the main purposes of this contribution is to combat any perception that organocatalysis is more a vehicle for chemists primarily interested in methodological development, while being less applicable to everyday natural product synthesis challenges. It has been shown clearly that this is no longer true. Furthermore, as the focus on the development of new modes of catalysis in general will become even more important in the future, the use of small-organic molecular catalysts should contribute substantially to render complex synthesis more efficient and straightforward.

References

1. Bhat SV, Nagasampagi BA, Sivakumar M (2005) Chemistry of Natural Products. Narosa Publishing House, New Delhi
2. Nicolaou KC, Sorensen EJ (1996) Classics in Total Synthesis: Targets, Strategies, Methods. VCH, Weinheim
3. Nicolaou KC, Snyder SA (2003) Classics in Total Synthesis II: More Targets, Strategies, Methods. Wiley-VCH, Weinheim
4. Nicolaou KC, Chen JS (2011) Classics in Total Synthesis III: Further Targets, Strategies, Methods. Wiley-VCH, Weinheim
5. Berkessel A, Gröger H (2005) Asymmetric Organocatalysis: From Biomimetic Concepts to Applications in Asymmetric Synthesis. Wiley-VCH, Weinheim
6. Dalko I (2007) Enantioselective Organocatalysis. Wiley-VCH, Weinheim
7. Gaunt MJ, Johansson CCC, McNally A, Vo NT (2007) Enantioselective Organocatalysis. Drug Discov Today **12**: 8
8. Seayad J, List B (2005) Asymmetric Organocatalysis. Org Biomol Chem **3**: 719
9. Bertelsen S, Jørgensen KA (2009) Organocatalysis – After the Gold Rush. Chem Soc Rev **38**: 2178
10. Breding G, Fiske PS (1912) Durch Katalysatoren Bewirkte Asymmetrische Synthese Biochem Z **46**: 7
11. Langenbeck W (1932) Enzyme Problems and Organic Catalysis. Angew Chem **45**: 97
12. Eder U, Sauer G, Weichert R (1971) New Type of Asymmetric Cyclization to Optically Active Steroid CD Partial Structures. Angew Chem Int Ed **10**: 496
13. Hajos ZG, Parrish DR (1974) Asymmetric Synthesis of Bicyclic Intermediates of Natural Product Chemistry. J Org Chem **39**: 1615
14. Hajos ZG, Parrish DR (1971) Asymmetric Synthesis of Optically Active Polycyclic Organic Compounds. German patent DE 2102623
15. Prelog V, Wilhelm M (1954) Untersuchungen über Asymmetrische Synthesen, 6. Der Reaktionsmechanismus und der Sterische Verlauf der Asymmetrischen Cyanhydrin-Synthese. Helv Chim Acta **37**: 1634
16. Pracejus H (1960) Organische Katalysatoren, LXI. Asymmetrische Synthesen mit Ketenen, 1. Alkaloid-Katalysierte Asymmetrische Synthesen von α-Phenyl-Propionsäureestern. Justus Liebigs Ann Chem **634**: 9
17. Pracejus H (1960) Asymmetrische Synthesen mit Ketenen, 2. Stereospezifische Addition von α-Phenyl-Äthylamin an Phenyl-Methyl-Keten. Justus Liebigs Ann Chem **634**: 23
18. Stork G, Terrell R, Szmuszkovicz J (1954) A New Synthesis of 2-Alkyl and 2-Acyl Ketones. J Am Chem Soc **76**: 2029

M. Waser, *Asymmetric Organocatalysis in Natural Product Syntheses*,
Progress in the Chemistry of Organic Natural Products 96,
DOI 10.1007/978-3-7091-1163-5, © Springer-Verlag Wien 2012

19. Dolling UH, Davis P, Grabowski EJJ (1984) Efficient Catalytic Asymmetric Alkylations. 1. Enantioselective Synthesis of (+)-Indacrinone *via* Chiral Phase-Transfer Catalysis. J Am Chem Soc **106**: 446

20. Kagan HB, Riant O (1992) Catalytic Asymmetric *Diels-Alder* Reactions. Chem Rev **92**: 1007

21. Nielsen M, Worgull D, Zweifel T, Gschwend B, Bertelsen S, Jørgensen KA (2011) Mechanisms in Aminocatalysis. Chem Commun **47**: 632

22. MacMillan DWC (2008) The Advent and Development of Organocatalysis. Nature **455**: 304

23. Nielsen M, Jacobsen CB, Holub N, Paixao MW, Jørgensen KA (2010) Asymmetric Organocatalysis with Sulfones. Angew Chem Int Ed **49**: 2668

24. Dondoni A, Massi A (2008) Asymmetric Organocatalysis: From Infancy to Adolescence. Angew Chem Int Ed **47**: 4638

25. Marcelli T, Hiemstra H (2010) *Cinchona* Alkaloids in Asymmetric Organocatalysis. Synthesis 1229

26. List B, Yang JW (2006) The Organic Approach to Asymmetric Catalysis. Science **313**: 1584

27. Ahrendt KA, Borths CJ, MacMillan DWC (2000) New Strategies for Organic Catalysis: The First Highly Enantioselective Organocatalytic *Diels-Alder* Reaction. J Am Chem Soc **122**: 4243

28. List B, Lerner RA, Barbas CF III (2000) Proline-Catalyzed Direct Asymmetric Aldol Reactions. J Am Chem Soc **122**: 2395

29. de Figueiredo RM, Christmann M (2007) Organocatalytic Synthesis of Drugs and Bioactive Natural Products. Eur J Org Chem 2575

30. Marques-Lopez E, Herrera RP, Christmann M (2010) Asymmetric Organocatalysis in Total Synthesis – A Trial by Fire. Nat Prod Rep **27**: 1138

31. Kazmaier U (2005) Amino Acids – Valuable Organocatalysts in Carbohydrate Synthesis. Angew Chem Int Ed **44**: 2186

32. Grondal C, Jeanty M, Enders D (2010) Organocatalytic Cascade Reactions as a New Tool in Total Synthesis. Nat Chem **2**: 167

33. Markert M, Mahrwald R (2008) Total Syntheses of Carbohydrates: Organocatalyzed Aldol Additions of Dihydroxyacetone. Chem Eur J **14**: 40

34. Hoang L, Bahmanyar S, Houk KN, List B (2003) Kinetic and Stereochemical Evidence for the Involvement of Only One Proline Molecule in the Transition States of Proline-Catalyzed Intra- and Intermolecular Aldol Reactions. J Am Chem Soc **125**: 16

35. List B (2004) Enamine Catalysis Is a Powerful Strategy for the Catalytic Generation and Use of Carbanion Equivalents. Acc Chem Res **37**: 548

36. Mukherjee S, Yang JW, Hoffmann S, List B (2007) Asymmetric Enamine Catalysis. Chem Rev **107**: 5471

37. Heathcock CH (1991) The Aldol Reaction. Acid and General Base Catalysis. In: Trost BM, Fleming I, Heathcock CH (eds) Comprehensive Organic Synthesis, vol 2. Pergamon, Oxford, p 133

38. Mahrwald R (2004) Modern Aldol Reactions. Wiley-VCH, Weinheim

39. Machajewski TD, Wong CH (2000) The Catalytic Asymmetric Aldol Reaction. Angew Chem Int Ed **39**: 1352

40. Johnson JS, Evans DA (2000) Chiral Bis(oxazoline) Copper(II) Complexes: Versatile Catalysts for Enantioselective Cycloaddition, Aldol, *Michael*, and Carbonyl Ene Reactions. Acc Chem Res **33**: 325

41. Denmark SE, Stavenger RA (2000) Asymmetric Catalysis of Aldol Reactions with Chiral *Lewis* Bases. Acc Chem Res **33**: 432

42. Alcaide B, Almendros P (2002) The Direct Catalytic Asymmetric Aldol Reaction. Eur J Org Chem 1595

43. Alcaide B, Almendros P (2003) The Direct Catalytic Asymmetric Cross-Aldol Reaction of Aldehydes. Angew Chem Int Ed **42**: 858

44. List B, Pojarliev P, Castello C (2001) Proline-Catalyzed Asymmetric Aldol Reactions Between Ketones and α-Unsubstituted Aldehydes. Org Lett **3**: 573

45. Ikishima H, Sekiguchi Y, Ichikawa Y, Kotsuki H (2006) Synthesis of (–)-(5*R*,6*S*)-6-Acetoxyhexadecanolide Based on L-Proline-Catalyzed Asymmetric Aldol Reactions. Tetrahedron **62**: 311

46. Sun B, Peng LZ, Chen XS, Li YL, Li Y, Yamasaki K (2005) Synthesis of (–)-(5*R*,6*S*)-6-Acetoxyhexadecan-5-olide by L-Proline-Catalyzed Asymmetric Aldol Reactions. Tetrahedron Asymmetry **16**: 1305
47. Goodin S, Kane MP, Rubin EH (2004) Epothilones: Mechanism of Action and Biologic Activity. J Clin Oncol **22**: 2015
48. Mulzer J, Prantz K (2009) Total Synthesis of Epothilones A-F. Prog Chem Org Nat Prod **90**: 5
49. Zheng Y, Avery MA (2004) Asymmetric Aldol Reactions Using Catalytic D-(+)-Proline: a New, Economic and Practical Approach to a Commonly Employed C1–C6 Keto-acid Synthon of the Epothilones. Tetrahedron **60**: 2091
50. Doi T, Numajiri Y, Munakata A, Takahashi T (2006) Total Synthesis of Apratoxin A. Org Lett **8**: 531
51. Luppi G, Monari M, Correa RJ, Violante FD, Pinto AC, Kaptein B, Broxterman QB, Garden SJ, Tomasini C (2006) The First Total Synthesis of (*R*)-Convolutamydine A. Tetrahedron **62**: 12017
52. Chen J-R, Liu X-P, Zhu X-Y, Qiao Y-F, Zhang J-M, Xiao W-J (2007) Organocatalytic Asymmetric Aldol Reaction of Ketones with Isatins: Straightforward Stereoselective Synthesis of 3-Alkyl-3-hydroxyindolin-2-ones. Tetrahedron **63**: 10437
53. Malkov AV, Kabeshov MA, Bella M, Kysilka O, Malyshev DA, Pluhackova K, Kocovsky P (2007) Vicinal Amino Alcohols as Organocatalysts in Asymmetric Cross-Aldol Reaction of Ketones: Application in the Synthesis of Convolutamydine A. Org Lett **9**: 5473
54. Nakamura S, Hara N, Nakashima H, Kubo K, Shibata N, Toru T (2008) Enantioselective Synthesis of (*R*)-Convolutamydine A with New *N*-Heteroarylsulfonylprolinamides. Chem Eur J **14**: 8079
55. Zhang HP, Shigemori H, Ishibashi M, Kosaka T, Pettit GR, Kamano Y, Kobayashi J (1994) Convolutamides A-F, Novel γ-Lactam Alkaloids from the Marine Bryozoan *Amathia convoluta*. Tetrahedron **50**: 10201
56. Kamano Y, Zhang HP, Ichihara Y, Kizu H, Komiyama K, Pettit GR (1995) Convolutamydine-A, a Novel Bioactive Hydroxyoxindole Alkaloid from Marine Bryozoan *Amathia convoluta*. Tetrahedron Lett **36**: 2783
57. Enders D, Grondal C (2005) Direct Organocatalytic de Novo Synthesis of Carbohydrates. Angew Chem Int Ed **44**: 1210
58. Enders D, Palecek J, Grondal C (2006) A Direct Organocatalytic Entry to Sphingoids: Asymmetric Synthesis of D-Arabino- and L-Ribo-Phytosphingosine. Chem Commun 655
59. Enders D, Voith M, Lenzen A (2005) The Dihydroxyacetone Unit – a Versatile C-3 Building Block in Organic Synthesis. Angew Chem Int Ed **44**: 1304
60. Enders D, Gasperi T (2007) Proline Organocatalysis as a New Tool for the Asymmetric Synthesis of Ulosonic Acid Precursors. Chem Commun 88
61. Entner N, Doudoroff M (1952) Glucose and Gluconic Acid Oxidation of *Pseudomonas saccharophila*. J Biol Chem **196**: 853
62. Ward DE, Jheengut V, Beye GE (2006) Thiopyran Route to Polypropionates: An Efficient Synthesis of Serricornin. J Org Chem **71**: 8989
63. Hartikka A, Arvidsson PI (2004) Rational Design of Asymmetric Organocatalysts – Increased Reactivity and Solvent Scope with a Tetrazolic Acid. Tetrahedron Asymmetry **15**: 1831
64. Hartikka A, Arvidsson PI (2005) 5-(Pyrrolidine-2-yl)-tetrazole: Rationale for the Increased Reactivity of the Tetrazole Analogue of Proline in Organocatalyzed Aldol Reactions. Eur J Org Chem 4287
65. Torii H, Nakadai M, Ishihara K, Saito S, Yamamoto H (2004) Asymmetric Direct Aldol Reaction Assisted by Water and a Proline-Derived Tetrazole Catalyst. Angew Chem Int Ed **43**: 1983
66. Cobb AJA, Shaw DM, Ley SV (2004) 5-Pyrrolidin-2-yltetrazole: A New, Catalytic, More Soluble Alternative to Proline in an Organocatalytic Asymmetric *Mannich*-Type Reaction. Synlett 558
67. Kumar I, Rode CV (2007) L-Proline Catalyzed Direct Diastereoselective Aldol Reactions: Towards the Synthesis of Lyxo-(2*S*,3*S*,4*S*)-phytosphingosine. Tetrahedron Asymmetry **18**: 1975

68. Sato Y, Fukuda H, Tomizawa M, Masaki T, Shibuya M, Kanoh N, Iwabuchi Y (2010) An Enantio- and Diastereocontrolled Synthesis of (–)-Salinosporamide A. Heterocycles **81**: 2239
69. Northrup AB, MacMillan DWC (2002) The First Direct and Enantioselective Cross-Aldol Reaction of Aldehydes. J Am Chem Soc **124**: 6798
70. Pihko PM, Erkkila A (2003) Enantioselective Synthesis of Prelactone B Using a Proline-Catalyzed Crossed-Aldol Reaction. Tetrahedron Lett **44**: 7607
71. Northrup AB, MacMillan DWC (2004) Two-Step Synthesis of Carbohydrates by Selective Aldol Reactions. Science **305**: 1752
72. Northrup AB, Mangion IK, Hettche F, MacMillan DWC (2004) Enantioselective Organocatalytic Direct Aldol Reactions of α-Oxyaldehydes: Step One in a Two-Step Synthesis of Carbohydrates. Angew Chem Int Ed **43**: 2152
73. Pettit GR, Cichacz ZA, Gao F, Herald CL, Boyd MR, Schmidt JM, Hooper JNA (1993) Antineoplastic Agents. 257. Isolation and Structure of Spongistatin-1. J Org Chem **58**: 1302
74. Bai RL, Cichacz ZA, Herald CL, Pettit GR, Hamel E (1993) Spongistatin-1, a Highly Cytotoxic, Sponge-Derived, Marine Natural Product that Inhibits Mitosis, Microtubule Assembly, and the Binding of Vinblastine to Tubulin. Mol Pharmacol **44**: 757
75. Pettit GR, Cichacz ZA, Gao F, Herald CL, Boyd MR (1993) Isolation and Structure of the Remarkable Human Cancer Cell-Growth Inhibitors Spongistatin-2 and Spongistatin-3 from an Eastern Indian Ocean-*Spongia* sp. J Chem Soc, Chem Commun 1166
76. Smith AB, III, Beauchamp TJ, LaMarche MJ, Kaufman MD, Qiu YP, Arimoto H, Jones DR, Kobayashi K (2000) Evolution of a Gram-Scale Synthesis of (+)-Discodermolide. J Am Chem Soc **122**: 8654
77. Smith AB, III, Lin QY, Doughty VA, Zhuang LH, McBriar MD, Kerns JK, Boldi AM, Murase N, Moser WH, Brook CS, Bennett CS, Nakayama K, Sobukawa M, Trout REL (2009) Spongipyran Synthetic Studies. Total Synthesis of (+)-Spongistatin 2. Tetrahedron **65**: 6470
78. Smith AB, III, Sfouggatakis C, Risatti CA, Sperry JB, Zhu WY, Doughty VA, Tomioka T, Gotchev DB, Bennett CS, Sakamoto S, Atasoylu O, Shirakami S, Bauer D, Takeuchi M, Koyanagi J, Sakamoto Y (2009) Spongipyran Synthetic Studies. Evolution of a Scalable Total Synthesis of (+)-Spongistatin 1. Tetrahedron **65**: 6489
79. Zhang SL, Duan WH, Wang W (2006) Efficient, Enantioselective Organocatalytic Synthesis of Trichostatin A. Adv Synth Catal **348**: 1228
80. Nakamura T, Shirokawa S, Hosokawa S, Nakazaki A, Kobayashi S (2006) Enantioselective Total Synthesis of Convolutamydines B and E. Org Lett **8**: 677
81. Hara N, Nakamura S, Shibata N, Toru T (2009) First Enantioselective Synthesis of (*R*)-Convolutamydine B and E with *N*-(Heteroarenesulfonyl)prolinamides. Chem Eur J **15**: 6790
82. Itoh T, Ishikawa H, Hayashi Y (2009) Asymmetric Aldol Reaction of Acetaldehyde and Isatin Derivatives for the Total Syntheses of *ent*-Convolutamydine E and CPC-1 and a Half Fragment of Madindoline A and B. Org Lett **11**: 3854
83. Kitajima M, Mori I, Arai K, Kogure N, Takayama H (2006) Two New Tryptamine-Derived Alkaloids from *Chimonanthus praecox* L. *concolor*. Tetrahedron Lett **47**: 3199
84. Hayashi M, Rho M-C, Enomoto A, Fukami A, Kim YP, Kikuchi Y, Sunazuka T, Hirose T, Komiyama K, Omura S (2002) Suppression of Bone Resorption by Madindoline A, a Novel Nonpeptide Antagonist to gp130. Proc Natl Acad Sci USA **99**: 14728
85. Pidathala C, Hoang L, Vignola N, List B (2003) Direct Catalytic Asymmetric Enolexo Aldolizations. Angew Chem Int Ed **42**: 2785
86. Mans DM, Pearson WH (2004) Total Synthesis of (+)-Cocaine *via* Desymmetrization of a *meso*-Dialdehyde. Org Lett **6**: 3305
87. Itagaki N, Kimura M, Sugahara T, Iwabuchi Y (2005) Organocatalytic Entry to Chiral Bicyclo[3.n.1]alkanones *via* Direct Asymmetric Intramolecular Aldolization. Org Lett **7**: 4185
88. Itagaki N, Sugahara T, Iwabuchi Y (2005) Expedient Synthesis of Potent Cannabinoid Receptor Agonist (–)-CP55,940. Org Lett **7**: 4181

89. Itagaki N, Iwabuchi Y (2007) Enantio- and Diastereocontrolled Synthesis of (+)-Juvabione Employing Organocatalytic Desymmetrisation and Photoinduced Fragmentation. Chem Commun 1175

90. Heathcock CH, Graham SL, Pirring MS, Playac F, White CT (1983) The Total Synthesis of Sesquiterpenes, 1970–1979. In: ApSimon J (ed) Total Synthesis of Natural Products, vol 5. Wiley, New York, p 1

91. Sarkar SM, Taira Y, Nakano A, Takahashi K, Ishihara J, Hatakeyama S (2011) Organocatalytic Asymmetric Synthesis of Quinine and Quinidine. Tetrahedron Lett **52**: 923

92. Cordova A (2004) The Direct Catalytic Asymmetric *Mannich* Reaction. Acc Chem Res **37**: 102

93. List B (2000) The Direct Catalytic Asymmetric Three-Component *Mannich* Reaction. J Am Chem Soc **122**: 9336

94. Nagata K, Nishimura K, Yokoya M, Itoh T (2006) Enantioselective Syntheses of *ent*-Sedridine and (+)-Coniine *via* Proline-Catalyzed *Mannich* Reaction. Heterocycles **70**: 335

95. Itoh T, Nishimura K, Nagata K, Yokoya M (2006) Total Synthesis of *ent*-Sedridine Using Proline-Catalyzed Asymmetric Addition as a Key Step. Synlett 2207

96. Hayashi Y, Urushima T, Shin M, Shoji M (2005) The Stereoselective Synthesis of α-Substituted β-Amino Secondary Alcohols Based on the Proline-Mediated, Asymmetric, Three-Component *Mannich* Reaction and its Application to the Formal Total Synthesis of Nikkomycins B and Bx. Tetrahedron **61**: 11393

97. Enders D, Vrettou M (2006) Asymmetric Synthesis of (+)-Polyoxamic Acid *via* an Efficient Organocatalytic *Mannich* Reaction as the Key Step. Synthesis 2155

98. Isono K, Asahi K, Suzuki S (1969) Studies on Polyoxins, Antifungal Antibiotics. 13. Structure of Polyoxins. J Am Chem Soc **91**: 7490

99. Paraskar AS, Sudalai A (2006) Enantioselective Synthesis of (–)-Cytoxazone and (+)-*epi*-Cytoxazone, Novel Cytokine Modulators *via Sharpless* Asymmetric Epoxidation and L-Proline Catalyzed *Mannich* Reaction. Tetrahedron **62**: 5756

100. Kakeya H, Morishita M, Kobinata K, Osono M, Ishizuka M, Osada H (1998) Isolation and Biological Activity of a Novel Cytokine Modulator, Cytoxazone. J Antibiot **51**: 1126

101. Enders D, Grondal C, Vrettou M, Raabe G (2005) Asymmetric Synthesis of Selectively Protected Amino Sugars and Derivatives by a Direct Organocatalytic *Mannich* Reaction. Angew Chem Int Ed **44**: 4079

102. Knapp S, Kukkola PJ, Sharma S, Dhar TGM, Naughton ABJ (1990) Amino Alcohol and Amino Sugar Synthesis by Benzoylcarbamate Cyclization. J Org Chem **55**: 5700

103. Winterfeld GA, Schmidt RR (2001) Nitroglycal Concatenation: A Broadly Applicable and Efficient Approach to the Synthesis of Complex O-Glycans. Angew Chem Int Ed **40**: 2654

104. Evans DA, Hu E, Tedrow JS (2001) An Aldol-Based Approach to the Asymmetric Synthesis of L-Callipeltose, the Deoxyamino Sugar of L-Callipeltoside A. Org Lett **3**: 3133

105. Guillena G, Ramon DJ (2006) Enantioselective α-Heterofunctionalization of Carbonyl Compounds: Organocatalysis Is the Simplest Approach. Tetrahedron Asymmetry **17**: 1465

106. Franzen J, Marigo M, Fielenbach D, Wabnitz TC, Kjaersgaard A, Jørgensen KA (2005) A General Organocatalyst for Direct α-Functionalization of Aldehydes: Stereoselective C–C, C–N, C–F, C–Br, and C–S Bond-Forming Reactions. Scope and Mechanistic Insights. J Am Chem Soc **127**: 18296

107. Hayashi Y, Yamaguchi J, Hibino K, Shoji M (2003) Direct Proline Catalyzed Asymmetric α-Aminooxylation of Aldehydes. Tetrahedron Lett **44**: 8293

108. Brown SP, Brochu MP, Sinz CJ, MacMillan DWC (2003) The Direct and Enantioselective Organocatalytic α-Oxidation of Aldehydes. J Am Chem Soc **125**: 10808

109. Harbindu A, Kumar P (2011) Organocatalytic Enantioselective Approach to the Synthesis of Verbalactone and (*R*)-Massoialactone. Synthesis 1954

110. Momiyama N, Yamamoto H (2003) Catalytic Enantioselective Synthesis of α-Aminooxy and α-Hydroxy Ketone Using Nitrosobenzene. J Am Chem Soc **125**: 6038

111. Kim SG, Park TH, Kim BJ (2006) Efficient Total Synthesis of (+)-Exo-, (–)-Endo-Brevicomin and Their Derivatives *via* Asymmetric Organocatalysis and Olefin Cross-Metathesis. Tetrahedron Lett **47**: 6369
112. Yamaguchi J, Toyoshima M, Shoji M, Kakeya H, Osada H, Hayashi Y (2006) Concise Enantio- and Diastereoselective Total Syntheses of Fumagillol, RK-805, FR65814, Ovalicin, and 5-Demethylovalicin. Angew Chem Int Ed **45**: 789
113. Matsuzawa M, Kakeya H, Yamaguchi J, Shoji M, Onose R, Osada H, Hayashi Y (2006) Enantio- and Diastereoselective Total Synthesis of (+)-Panepophenanthrin, a Ubiquitin-Activating Enzyme Inhibitor, and Biological Properties of its New Derivatives. Chem Asian J **1**: 845
114. Sekizawa R, Ikeno S, Nakamura H, Naganawa H, Matsui S, Iinuma H, Takeuchi T (2002) Panepophenanthrin, from a Mushroom Strain, a Novel Inhibitor of the Ubiquitin-Activating Enzyme. J Nat Prod **65**: 1491
115. Mangion IK, MacMillan DWC (2005) Total Synthesis of Brasoside and Littoralisone. J Am Chem Soc **127**: 3696
116. Varseev GN, Maier ME (2007) Enantioselective Total Synthesis of (+)-Neosymbioimine. Org Lett **9**: 1461
117. Kita M, Ohishi N, Washida K, Kondo M, Koyama T, Yamada K, Uemura D (2005) Symbioimine and Neosymbioimine, Amphoteric Iminium Metabolites from the Symbiotic Marine Dinoflagellate *Symbiodinium* sp. Bioorg Med Chem **13**: 5253
118. Bierl BA, Beroza M, Collier CW (1970) Potent Sex Attractant of Gypsy Moth – Its Isolation, Identification, and Synthesis. Science **170**: 87
119. Miller JR, Mori K, Roelofs WL (1977) Gypsy Moth Field Trapping and Electroantennogram Studies with Pheromone Enantiomers. J Insect Physiol **23**: 1447
120. Kim SG (2009) Concise Total Synthesis of (+)-Disparlure and its *trans*-Isomer Using Asymmetric Organocatalysis. Synthesis 2418
121. Park Y, Tae J (2010) Facile Synthesis of (–)-6-Acetoxy-5-hexadecanolide by Organocatalytic α-Oxygenation-Allylation-RCM Strategy. Synthesis 3627
122. Kano T, Mii H, Maruoka K (2009) Direct Asymmetric Benzoyloxylation of Aldehydes Catalyzed by 2-Tritylpyrrolidine. J Am Chem Soc **131**: 3450
123. Bogevig A, Juhl K, Kumaragurubaran N, Zhuang W, Jørgensen KA (2002) Direct Organocatalytic Asymmetric α-Amination of Aldehydes – A Simple Approach to Optically Active α-Amino Aldehydes, α-Amino Alcohols, and α-Amino Acids. Angew Chem Int Ed **41**: 1790
124. Kumaragurubaran N, Juhl K, Zhuang W, Bogevig A, Jørgensen KA (2002) Direct L-Proline-Catalyzed Asymmetric α-Amination of Ketones. J Am Chem Soc **124**: 6254
125. List B (2002) Direct Catalytic Asymmetric α-Amination of Aldehydes. J Am Chem Soc **124**: 5656
126. Chowdari NS, Ramachary DB, Barbas CF (2003) Organocatalytic Asymmetric Assembly Reactions: One-Pot Synthesis of Functionalized β-Amino Alcohols from Aldehydes, Ketones, and Azodicarboxylates. Org Lett **5**: 1685
127. Chowdari NS, Barbas CF (2005) Total Synthesis of LFA-1 Antagonist BIRT-377 *via* Organocatalytic Asymmetric Construction of a Quaternary Stereocenter. Org Lett **7**: 867
128. Henmi Y, Makino K, Yoshitomi Y, Hara O, Hamada Y (2004) Highly Efficient Synthesis of (*R*)- and (*S*)-Piperazic Acids Using Proline-Catalyzed Asymmetric α-Hydrazination. Tetrahedron Asymmetry **15**: 3477
129. Ma XQ, Gang DR (2004) The *Lycopodium* Alkaloids. Nat Prod Rep **21**: 752
130. Yang H, Carter RG, Zakharov LN (2008) Enantioselective Total Synthesis of Lycopodine. J Am Chem Soc **130**: 9238
131. Bisai A, West SP, Sarpong R (2008) Unified Strategy for the Synthesis of the "Miscellaneous" *Lycopodium* Alkaloids: Total Synthesis of (+/−)-Lyconadin A. J Am Chem Soc **130**: 7222
132. Kozak JA, Dake GR (2008) Total Synthesis of (+)-Fawcettidine. Angew Chem Int Ed **47**: 4221

133. Linghu X, Kennedy-Smith JJ, Toste FD (2007) Total Synthesis of (+)-Fawcettimine. Angew Chem Int Ed **46**: 7671

134. Snider BB, Grabowski JF (2007) Total Synthesis of (–)-Senepodine G and (–)-Cermizine C. J Org Chem **72**: 1039

135. Yang H, Carter RG (2010) Development of an Enantioselective Route Toward the *Lycopodium* Alkaloids: Total Synthesis of Lycopodine. J Org Chem **75**: 4929

136. Morita H, Hirasawa Y, Shinzato T, Kobayashi J (2004) New Phlegmarane-Type, Cernuane-Type, and Quinolizidine Alkaloids from Two Species of *Lycopodium*. Tetrahedron **60**: 7015

137. Marion L, Manske RHF (1948) The Alkaloids of *Lycopodium* Species 10. *Lycopodium cernuum* L. Can J Res **26**: 1

138. Nishikawa Y, Kitajima M, Takayama H (2008) First Asymmetric Total Syntheses of Cernuane-Type *Lycopodium* Alkaloids, Cernuine, and Cermizine D. Org Lett **10**: 1987

139. Nishikawa Y, Kitajima M, Kogure N, Takayama H (2009) A Divergent Approach for the Total Syntheses of Cernuane-Type and Quinolizidine-Type *Lycopodium* Alkaloids. Tetrahedron **65**: 1608

140. Shigeyama T, Katakawa K, Kogure N, Kitajima M, Takayama H (2007) Asymmetric Total Syntheses of Two Phlegmarine-Type Alkaloids, Lycoposerramines-V and -W, Newly Isolated from *Lycopodium serratum*. Org Lett **9**: 4069

141. Katakawa K, Nozoe A, Kogure N, Kitajima M, Hosokawa M, Takayama H (2007) Fawcettimine-Related Alkaloids from *Lycopodium serratum*. J Nat Prod **70**: 1024

142. Katakawa K, Kitajima M, Aimi N, Seki H, Yamaguchi K, Furihata K, Harayama T, Takayama H (2005) Structure Elucidation and Synthesis of Lycoposerramine-B, a Novel Oxime-Containing *Lycopodium* Alkaloid from *Lycopodium serratum* Thunb. J Org Chem **70**: 658

143. Betancort JM, Sakthivel K, Thayumanavan R, Barbas CF (2001) Catalytic Enantioselective Direct *Michael* Additions of Ketones to Alkylidene Malonates. Tetrahedron Lett **42**: 4441

144. Sakthivel K, Notz W, Bui T, Barbas CF (2001) Amino Acid Catalyzed Direct Asymmetric Aldol Reactions: A Bioorganic Approach to Catalytic Asymmetric Carbon-Carbon Bond-Forming Reactions. J Am Chem Soc **123**: 5260

145. Melchiorre P, Jørgensen KA (2003) Direct Enantioselective *Michael* Addition of Aldehydes to Vinyl Ketones Catalyzed by Chiral Amines. J Org Chem **68**: 4151

146. Fonseca MTH, List B (2004) Catalytic Asymmetric Intramolecular *Michael* Reaction of Aldehydes. Angew Chem Int Ed **43**: 3958

147. Peelen TJ, Chi YG, Gellman SH (2005) Enantioselective Organocatalytic *Michael* Additions of Aldehydes to Enones with Imidazolidinones: Cocatalyst Effects and Evidence for an Enamine Intermediate. J Am Chem Soc **127**: 11598

148. Hayashi Y, Gotoh H, Tamura T, Yamaguchi H, Masui R, Shoji M (2005) Cysteine-Derived Organocatalyst in a Highly Enantioselective Intramolecular *Michael* Reaction. J Am Chem Soc **127**: 16028

149. Andrey O, Alexakis A, Tomassini A, Bernardinelli G (2004) The Use of *N*-Alkyl-2,2′-Bipyrrolidine Derivatives as Organocatalysts for the Asymmetric *Michael* Addition of Ketones and Aldehydes to Nitroolefins. Adv Synth Catal **346**: 1147

150. Andrey O, Alexakis A, Bernardinelli G (2003) Asymmetric *Michael* Addition of α-Hydroxyketones to Nitroolefins Catalyzed by Chiral Diamine. Org Lett **5**: 2559

151. Ishikawa H, Suzuki T, Hayashi Y (2009) High-Yielding Synthesis of the Anti-influenza Neuramidase Inhibitor (–)-Oseltamivir by Three "One-Pot" Operations. Angew Chem Int Ed **48**: 1304

152. Kim CU, Lew W, Williams MA, Liu HT, Zhang LJ, Swaminathan S, Bischofberger N, Chen MS, Mendel DB, Tai CY, Laver WG, Stevens RC (1997) Influenza Neuraminidase Inhibitors Possessing a Novel Hydrophobic Interaction in the Enzyme Active Site: Design, Synthesis, and Structural Analysis of Carbocyclic Sialic Acid Analogues with Potent Anti-influenza Activity. J Am Chem Soc **119**: 681

153. Andrey O, Vidonne A, Alexakis A (2003) Organocatalytic *Michael* Addition, a Convenient Tool in Total Synthesis. First Asymmetric Synthesis of (–)-Botryodiplodin. Tetrahedron Lett **44**: 7901

154. Gupta RS, Chandran RR, Divekar PV (1966) Botryodiplodin – a New Antibiotic from *Botryodiplodia theobromae*. Part I. Production, Isolation and Biological Properties. Ind J Exp Biol **4**: 152
155. Nicolaou KC, Sarlah D, Shaw DM (2007) Total Synthesis and Revised Structure of Biyouyanagin A. Angew Chem Int Ed **46**: 4708
156. Tanaka N, Okasaka M, Ishimaru Y, Takaishi Y, Sato M, Okamoto M, Oshikawa T, Ahmed SU, Consentino LM, Lee KH (2005) Biyouyanagin A, an Anti-HIV Agent from *Hypericum chinense* L. var. *salicifolium*. Org Lett **7**: 2997
157. Chi YG, Gellman SH (2005) Diphenylprolinol Methyl Ether: A Highly Enantioselective Catalyst for *Michael* Addition of Aldehydes to Simple Enones. Org Lett **7**: 4253
158. Chen K, Baran PS (2009) Total Synthesis of Eudesmane Terpenes by Site-Selective C-H Oxidations. Nature **459**: 824
159. Wu QX, Shi YP, Jia ZJ (2006) Eudesmane Sesquiterpenoids from the Asteraceae Family. Nat Prod Rep **23**: 699
160. Campbell MJ, Johnson JS (2009) Asymmetric Synthesis of (+)-Polyanthellin A. J Am Chem Soc **131**: 10370
161. ElMarrouni A, Joolakanti SR, Colon A, Heras M, Arseniyadis S, Cossy J (2010) Two Concise Total Syntheses of (–)-Bitungolide F. Org Lett **12**: 4074
162. Knüppel S, Rogachev VO, Metz P (2010) A Concise Catalytic Route to the Marine Sesquiterpenoids (–)-Clavukerin A and (–)-Isoclavukerin A. Eur J Org Chem **2010**: 6145
163. Gloer JB, Rinderknecht BL, Wicklow DT, Dowd PF (1989) Nominine – a New Insecticidal Indole Diterpene from the Sclerotia of *Aspergillus nomius*. J Org Chem **54**: 2530
164. Bradshaw B, Etxebarria-Jardi G, Bonjoch J (2010) Total Synthesis of (–)-Anominine. J Am Chem Soc **132**: 5966
165. Bradshaw B, Etxebarria-Jardi G, Bonjoch J, Viozquez SF, Guillena G, Najera C (2009) Efficient Solvent-Free *Robinson* Annulation Protocols for the Highly Enantioselective Synthesis of the *Wieland-Miescher* Ketone and Analogues. Adv Synth Catal **351**: 2482
166. Bertelsen S, Marigo M, Brandes S, Diner P, Jørgensen KA (2006) Dienamine Catalysis: Organocatalytic Asymmetric γ-Amination of α,β–Unsaturated Aldehydes. J Am Chem Soc **128**: 12973
167. Hong B-C, Wu M-F, Tseng H-C, Huang G-F, Su C-F, Liao J-H (2007) Organocatalytic Asymmetric *Robinson* Annulation of α,β-Unsaturated Aldehydes: Applications to the Total Synthesis of (+)-Palitantin. J Org Chem **72**: 8459
168. Birkinshaw JH, Raistrick H (1936) Studies in the Biochemistry of Micro-organisms. XLIX. Palitantin, $C_{14}H_{22}O_4$, a Hitherto Undescribed Metabolic Product of *Penicillium palitans* Westling. Biochem J **30**: 801
169. Bowden K, Lythgoe B, Marsden DJS (1959) The Structure of Palitantin. J Chem Soc 1662
170. Hanessian S, Sakito Y, Dhanoa D, Baptistella L (1989) Synthesis of (+)-Palitantin. Tetrahedron **45**: 6623
171. Mahapatra T, Nanda S (2009) Asymmetric Synthesis of Palitantin by an Enzymatic and Organocatalytic Approach. Tetrahedron Asymmetry **20**: 610
172. de Figueiredo RM, Fröhlich R, Christmann M (2008) Amine-Catalyzed Cyclizations of Tethered α,β-Unsaturated Carbonyl Compounds. Angew Chem Int Ed **47**: 1450
173. Marques-Lopez E, Herrera RP, Marks T, Jacobs WC, Konning D, de Figueiredo RM, Christmann M (2009) Crossed Intramolecular *Rauhut-Currier*-Type Reactions *via* Dienamine Activation. Org Lett **11**: 4116
174. Starkenmann C, Mayenzet F, Brauchli R, Wunsche L, Vial C (2007) Structure Elucidation of a Pungent Compound in Black Cardamom: *Amomum tsao-ko* Crevost et Lemarie (Zingiberaceae). J Agric Food Chem **55**: 10902
175. Watanabe K, Takada Y, Matsuo N, Nishimura H (1995) Rotundial, a New Natural Mosquito Repellent from the Leaves of *Vitex rotundifolia*. Biosci Biotechnol Biochem **59**: 1979
176. Enders D, Hüttl MRM, Grondal C, Raabe G (2006) Control of Four Stereocentres in a Triple Cascade Organocatalytic Reaction. Nature **441**: 861

177. Alachraf MW, Handayani PP, Hüttl MRM, Grondal C, Enders D, Schrader W (2011) Electrospray Mass Spectrometry for Detailed Mechanistic Studies of a Complex Organocatalyzed Triple Cascade Reaction. Org Biomol Chem **9**: 1047
178. Ramachary DB, Jain S (2011) Sequential One-Pot Combination of Multi-component and Multi-catalysis Cascade Reactions: An Emerging Technology in Organic Synthesis. Org Biomol Chem **9**: 1277
179. Itoh T, Yokoya M, Miyauchi K, Nagata K, Ohsawa A (2006) Total Synthesis of *ent*-Dihydrocorynantheol by Using a Proline-Catalyzed Asymmetric Addition Reaction. Org Lett **8**: 1533
180. Gilbert B, Antonaccio LD, Djerassi C (1962) Alkaloid Studies. 39. Occurrence of Dihydrocorynantheol and Aricine in *Aspidosperma marcgravianum* Woodson. J Org Chem **27**: 4702
181. Li YS, Matsunaga K, Ishibashi M, Ohizumi Y (2001) Littoralisone, a Novel Neuritogenic Iridolactone Having an Unprecedented Heptacyclic Skeleton Including Four- and Nine-Membered Rings Consisting of Glucose from *Verbena littoralis*. J Org Chem **66**: 2165
182. Rimpler H, Schafer B (1979) Hastatoside, a New Iridoid from *Verbena hastata* L. and *Verbena officinalis* L. Z Naturforsch **34**: 311
183. Franke A, Rimpler H (1987) Vebraside, an Iridoid Glucoside from *Verbena brasiliensis*. Phytochemistry **26**: 3015
184. Carpenter J, Northrup AB, Chung D, Wiener JJM, Kim SG, MacMillan DWC (2008) Total Synthesis and Structural Revision of Callipeltoside C. Angew Chem Int Ed **47**: 3568
185. Zampella A, D'Auria MV, Minale L, Debitus C, Roussakis C (1996) Callipeltoside A: A Cytotoxic Aminodeoxy Sugar-Containing Macrolide of a New Type from the Marine Lithistida Sponge *Callipelta* sp. J Am Chem Soc **118**: 11085
186. Zampella A, D'Auria MV, Minale L, Debitus C (1997) Callipeltosides B and C, Two Novel Cytotoxic Glycoside Macrolides from a Marine Lithistida Sponge *Callipelta* sp. Tetrahedron **53**: 3243
187. Liu KG, Chougnet A, Woggon WD (2008) A Short Route to α-Tocopherol. Angew Chem Int Ed **47**: 5827
188. Baldwin JE, Chesworth R, Parker JS, Russell AT (1995) Studies Towards a Postulated Biomimetic *Diels-Alder* Reaction for the Synthesis of Himgravine. Tetrahedron Lett **36**: 9551
189. Tchabanenko K, Chesworth R, Parker JS, Anand NK, Russell AT, Adlington RM, Baldwin JE (2005) Biomimetic Approach to *Galbulimima* Type I Alkaloids. Tetrahedron **61**: 11649
190. Tchabanenko K, Adlington RM, Cowley AR, Baldwin JE (2005) Biomimetic Total Synthesis of (+)-Himbacine. Org Lett **7**: 585
191. Kinsman AC, Kerr MA (2003) The Total Synthesis of (+)-Hapalindole Q by an Organomediated *Diels-Alder* Reaction. J Am Chem Soc **125**: 14120
192. Moore RE, Cheuk C, Patterson GML (1984) Hapalindoles – New Alkaloids from the Blue-Green Alga *Hapalosiphon fontinalis*. J Am Chem Soc **106**: 6456
193. Moore RE, Cheuk C, Yang XQG, Patterson GML, Bonjouklian R, Smitka TA, Mynderse JS, Foster RS, Jones ND, Swartzendruber JK, Deeter JB (1987) Hapalindoles, Antibacterial and Antimycotic Alkaloids from the Cyanophyte *Hapalosiphon fontinalis*. J Org Chem **52**: 1036
194. Jones SB, Simmons B, MacMillan DWC (2009) Nine-Step Enantioselective Total Synthesis of (+)-Minfiensine. J Am Chem Soc **131**: 13606
195. Wilson RM, Jen WS, MacMillan DWC (2005) Enantioselective Organocatalytic Intramolecular *Diels-Alder* Reactions. The Asymmetric Synthesis of Solanapyrone D. J Am Chem Soc **127**: 11616
196. Selkälä SA, Koskinen AMP (2005) Preparation of Bicyclo[4.3.0]nonanes by an Organocatalytic Intramolecular *Diels-Alder* Reaction. Eur J Org Chem 1620
197. Kumpulainen ETT, Koskinen AMP, Rissanen K (2007) Total Synthesis of Amaminol A: Establishment of the Absolute Stereochemistry. Org Lett **9**: 5043
198. Jacobs WC, Christmann M (2008) Synthesis of Amaminol B. Synlett 247

199. Oikawa H, Yokota T, Ichihara A, Sakamura S (1989) Structure and Absolute Configuration of Solanapyrone-D – A New Clue to the Occurrence of Biological *Diels-Alder* Reactions. J Chem Soc Chem Commun 1284
200. Oikawa H, Yokota T, Sakano C, Suzuki Y, Naya A, Ichihara A (1998) Solanapyrones, Phytotoxins Produced by *Alternaria solani*: Biosynthesis and Isolation of Minor Components. Biosci Biotechnol Biochem **62**: 2016
201. Hagiwara H, Kobayashi K, Miya S, Hoshi T, Suzuki T, Ando M, Okamoto T, Kobayashi M, Yamamoto I, Ohtsubo S, Kato M, Uda H (2002) First Total Syntheses of the Phytotoxins Solanapyrones D and E *via* the Domino *Michael* Protocol. J Org Chem **67**: 5969
202. Sata NU, Fusetani N (2000) Amaminols A and B, New Bicyclic Amino Alcohols from an Unidentified Tunicate of the Family Polyclinidae. Tetrahedron Lett **41**: 489
203. Yang JW, Fonseca MTH, List B (2004) A Metal-Free Transfer Hydrogenation: Organocatalytic Conjugate Reduction of α,β-Unsaturated Aldehydes. Angew Chem Int Ed **43**: 6660
204. Yang JW, Fonseca MTH, Vignola N, List B (2005) Metal-Free, Organocatalytic Asymmetric Transfer Hydrogenation of α,β-Unsaturated Aldehydes. Angew Chem Int Ed **44**: 108
205. Ouellet SG, Tuttle JB, MacMillan DWC (2005) Enantioselective Organocatalytic Hydride Reduction. J Am Chem Soc **127**: 32
206. Lacour J, Hebbe-Viton V (2003) Recent Developments in Chiral Anion Mediated Asymmetric Chemistry. Chem Soc Rev **32**: 373
207. Mukherjee S, List B (2007) Chiral Counteranions in Asymmetric Transition-Metal Catalysis: Highly Enantioselective Pd/*Brønsted* Acid-Catalyzed Direct α-Allylation of Aldehydes. J Am Chem Soc **129**: 11336
208. Hamilton GL, Kang EJ, Mba M, Toste FD (2007) A Powerful Chiral Counterion Strategy for Asymmetric Transition Metal Catalysis. Science **317**: 496
209. Mayer S, List B (2006) Asymmetric Counteranion-Directed Catalysis. Angew Chem Int Ed **45**: 4193
210. Martin NJA, List B (2006) Highly Enantioselective Transfer Hydrogenation of α,β-Unsaturated Ketones. J Am Chem Soc **128**: 13368
211. Li CQ, Wang C, Villa-Marcos B, Xiao JL (2008) Chiral Counteranion-Aided Asymmetric Hydrogenation of Acyclic Imines. J Am Chem Soc **130**: 14450
212. Wang X, List B (2008) Asymmetric Counteranion-Directed Catalysis for the Epoxidation of Enals. Angew Chem Int Ed **47**: 1119
213. Eey STC, Lear MJ (2010) A Bismuth(III)-Catalyzed *Friedel-Crafts* Cyclization and Stereocontrolled Organocatalytic Approach to (–)-Platensimycin. Org Lett **12**: 5510
214. Wang J, Soisson SM, Young K, Shoop W, Kodali S, Galgoci A, Painter R, Parthasarathy G, Tang YS, Cummings R, Ha S, Dorso K, Motyl M, Jayasuriya H, Ondeyka J, Herath K, Zhang CW, Hernandez L, Allocco J, Basilio A, Tormo JR, Genilloud O, Vicente F, Pelaez F, Colwell L, Lee SH, Michael B, Felcetto T, Gill C, Silver LL, Hermes JD, Bartizal K, Barrett J, Schmatz D, Becker JW, Cully D, Singh SB (2006) Platensimycin is a Selective FabF Inhibitor with Potent Antibiotic Properties. Nature **441**: 358
215. Singh SB, Jayasuriya H, Ondeyka JG, Herath KB, Zhang CW, Zink DL, Tsou NN, Ball RG, Basilio A, Genilloud O, Diez MT, Vicente F, Pelaez F, Young K, Wang J (2006) Isolation, Structure, and Absolute Stereochemistry of Platensimycin, a Broad Spectrum Antibiotic Discovered Using an Antisense Differential Sensitivity Strategy. J Am Chem Soc **128**: 11916
216. Nicolaou KC, Li A, Edmonds DJ (2006) Total Synthesis of Platensimycin. Angew Chem Int Ed **45**: 7086
217. Nicolaou KC, Edmonds DJ, Li A, Tria GS (2007) Asymmetric Total Syntheses of Platensimycin. Angew Chem Int Ed **46**: 3942
218. Lalic G, Corey EJ (2007) An Effective Enantioselective Route to the Platensimycin Core. Org Lett **9**: 4921
219. Li PF, Payette JN, Yamamoto H (2007) Enantioselective Route to Platensimycin: An Intramolecular *Robinson* Annulation Approach. J Am Chem Soc **129**: 9534

220. Tiefenbacher K, Mulzer J (2007) Protecting-Group-Free Formal Synthesis of Platensimycin. Angew Chem Int Ed **46**: 8074
221. Owusu-Ansah E, Durow AC, Harding JR, Jordan AC, O'Connell SJ, Willis CL (2011) Synthesis of Dysideaproline E Using Organocatalysis. Org Biomol Chem **9**: 265
222. Harrigan GG, Goetz GH, Luesch H, Yang ST, Likos J (2001) Dysideaprolines A-F and Barbaleucamides A-B, Novel Polychlorinated Compounds from a *Dysidea* Species. J Nat Prod **64**: 1133
223. Jung ME (1991) Stabilized Nucleophiles with Electron Deficient Alkenes and Alkynes. In: Trost BM, Fleming I (eds) Comprehensive Organic Synthesis, vol 4. Pergamon, Oxford, p 1
224. Ono N (2001) The Nitro Group in Organic Synthesis. Wiley-VCH, New York
225. Yamaguchi M (1999) Conjugate Addition of Stabilized Carbanions. In: Jacobsen EN, Pfaltz A, Yamamoto H (eds) Comprehensive Asymmetric Catalysis I–III, vol 3. Springer, Berlin, p 1121
226. Perlmutter P (1992) Conjugate Addition Reactions in Organic Synthesis. Pergamon, Oxford
227. Tsogoeva SB (2007) Recent Advances in Asymmetric Organocatalytic 1,4-Conjugate Additions. Eur J Org Chem 1701
228. Sibi MP, Manyem S (2000) Enantioselective Conjugate Additions. Tetrahedron **56**: 8033
229. Krause N, Hoffmann-Röder A (2001) Recent Advances in Catalytic Enantioselective *Michael* Additions. Synthesis 171
230. Halland N, Hansen T, Jørgensen KA (2003) Organocatalytic Asymmetric *Michael* Reaction of Cyclic 1,3-Dicarbonyl Compounds and α,β-Unsaturated Ketones – A Highly Atom-Economic Catalytic One-Step Formation of Optically Active Warfarin Anticoagulant. Angew Chem Int Ed **42**: 4955
231. Brandau S, Landa A, Franzen J, Marigo M, Jørgensen KA (2006) Organocatalytic Conjugate Addition of Malonates to α,β-Unsaturated Aldehydes: Asymmetric Formal Synthesis of (–)-Paroxetine, Chiral Lactams, and Lactones. Angew Chem Int Ed **45**: 4305
232. Valero G, Schimer J, Cisarova I, Vesely J, Moyano A, Rios R (2009) Highly Enantioselective Organocatalytic Synthesis of Piperidines. Formal Synthesis of (–)-Paroxetine. Tetrahedron Lett **50**: 1943
233. Overman RS, Stahmann MA, Huebner CF, Sullivan WR, Spero L, Doherty DG, Ikawa M, Graf L, Roseman S, Link KP (1944) Studies on the Hemorrhagic Sweet Clover Disease. XIII. Anticoagulant Activity and Structure in the 4-Hydroxycoumarin Group. J Biol Chem **153**: 5
234. O'Reilly RA (1976) The Stereoselective Interaction of Warfarin and Metronidazole in Man. N Engl J Med **295**: 354
235. Wingard LB, O'Reilly RA, Levy G (1978) Pharmacokinetics of Warfarin Enantiomers: A Search for Intrasubject Correlations. Clin Pharmacol Ther **23**: 212
236. Mukaiyama T, Banno K, Narasaka K (1974) New Cross-Aldol Reactions – Reactions of Silyl Enol Ethers with Carbonyl-Compounds Activated by Titanium Tetrachloride. J Am Chem Soc **96**: 7503
237. Mukaiyama T, Narasaka K, Banno K (1973) New Aldol Type Reaction. Chem Lett 1011
238. Nelson SG (1998) Catalyzed Enantioselective Aldol Additions of Latent Enolate Equivalents. Tetrahedron Asymmetry **9**: 357
239. Carreira EM (1999) Mukaiyama Aldol Reaction. In: Jacobsen EN, Pfaltz A, Yamamoto H (eds) Comprehensive Asymmetric Catalysis I–III, vol 3. Springer, Berlin, p 997
240. Carreira EM (2000) Aldol Reaction: Methodology and Stereochemistry. In: Otera J (ed) Modern Carbonyl Chemistry. Wiley, Weinheim, p 227
241. Evans DA, Murry JA, Kozlowski MC (1996) C_2-Symmetric Copper(II) Complexes as Chiral *Lewis* Acids. Catalytic Enantioselective Aldol Additions of Silylketene Acetals to (Benzyloxy)acetaldehyde. J Am Chem Soc **118**: 5814
242. Evans DA, Kozlowski MC, Burgey CS, MacMillan DWC (1997) C_2-Symmetric Copper(II) Complexes as Chiral *Lewis* Acids. Catalytic Enantioselective Aldol Additions of Enolsilanes to Pyruvate Esters. J Am Chem Soc **119**: 7893
243. Szlosek M, Figadere B (2000) Highly Enantioselective Aldol Reaction with 2-Trimethylsilyloxyfuran: The First Catalytic Asymmetric Autoinductive Aldol Reaction. Angew Chem Int Ed **39**: 1799

244. Brown SP, Goodwin NC, MacMillan DWC (2003) The First Enantioselective Organocatalytic *Mukaiyama-Michael* Reaction: A Direct Method for the Synthesis of Enantioenriched γ-Butenolide Architecture. J Am Chem Soc **125**: 1192
245. Robichaud J, Tremblay F (2006) Formal Enantioselective Synthesis of (+)-Compactin. Org Lett **8**: 597
246. Endo A, Kuroda M, Tsujita Y (1976) Ml-236a, Ml-236b, and Ml-236c, New Inhibitors of Cholesterogenesis Produced by *Penicillium citrinum*. J Antibiot **29**: 1346
247. Endo A, Kuroda M, Tanzawa K (1976) Competitive Inhibition of 3-Hydroxy-3-Methylglutaryl Coenzyme a Reductase by Ml-236a and Ml-236b Fungal Metabolites, Having Hypocholesterolemic Activity. FEBS Lett **72**: 323
248. Endo A, Tsujita Y, Kuroda M, Tanzawa K (1977) Inhibition of Cholesterol-Synthesis *in Vitro* and *in Vivo* by Ml-236a and Ml-236b, Competitive Inhibitors of 3-Hydroxy-3-Methylglutaryl-Coenzyme A Reductase. Eur J Biochem **77**: 31
249. Burnell RH (1959) *Lycopodium* Alkaloids. 1. Extraction of Alkaloids from *Lycopodium fawcettii*, Lloyd and Underwood. J Chem Soc 3091
250. Palomo C, Landa A, Mielgo A, Oiarbide M, Puente A, Vera S (2007) Water-Compatible Iminium Activation: Organocatalytic *Michael* Reactions of Carbon-Centered Nucleophiles with Enals. Angew Chem Int Ed **46**: 8431
251. Zu L, Xie H, Li H, Wang H, Wang W (2007) Highly Enantioselective Organocatalytic Conjugate Addition of Nitromethane to α,β-Unsaturated Aldehydes: Three-Step Synthesis of Optically Active Baclofen. Adv Synth Catal **349**: 2660
252. Szanto G, Hegedus L, Mattyasovszky L, Simon A, Simon A, Bitter I, Toth G, Toke L, Kadas I (2009) An Expedient Total Synthesis of *ent*-(–)-7-Deoxy-*trans*-Dihydronarciclasine. Tetrahedron **65**: 8412
253. Pettit GR, Pettit GR, Backhaus RA, Boyd MR, Meerow AW (1993) Antineoplastic Agents, 256. Cell-Growth Inhibitory Isocarbostyrils from *Hymenocallis*. J Nat Prod **56**: 1682
254. Paras NA, MacMillan DWC (2001) New Strategies in Organic Catalysis: The First Enantioselective Organocatalytic *Friedel-Crafts* Alkylation. J Am Chem Soc **123**: 4370
255. Paras NA, MacMillan DWC (2002) The Enantioselective Organocatalytic 1,4-Addition of Electron-Rich Benzenes to α,β-Unsaturated Aldehydes. J Am Chem Soc **124**: 7894
256. Austin JF, MacMillan DWC (2002) Enantioselective Organocatalytic Indole Alkylations. Design of a New and Highly Effective Chiral Amine for Iminium Catalysis. J Am Chem Soc **124**: 1172
257. Austin JF, Kim SG, Sinz CJ, Xiao WJ, MacMillan DWC (2004) Enantioselective Organocatalytic Construction of Pyrroloindolines by a Cascade Addition-Cyclization Strategy: Synthesis of (–)-Flustramine B. Proc Natl Acad Sci USA **101**: 5482
258. Carle JS, Christophersen C (1979) Bromo-Substituted Physostigmine Alkaloids from a Marine Bryozoa *Flustra foliacea*. J Am Chem Soc **101**: 4012
259. Carle JS, Christophersen C (1980) Marine Alkaloids. 2. Bromo Alkaloids from the Marine Bryozoan *Flustra foliacea* – Isolation and Structure Elucidation. J Org Chem **45**: 1586
260. Carle JS, Christophersen C (1981) Marine Alkaloids. 3. Bromo-Substituted Alkaloids from the Marine Bryozoan *Flustra foliacea*, Flustramine C and Flustraminol A and Flustraminol B. J Org Chem **46**: 3440
261. Banwell MG, Beck DAS, Willis AC (2006) Enantioselective Total Syntheses of the Alkaloids (–)-Rhazinal, (–)-Rhazinilam, (–)-Leuconolam and (+)-*epi*-Leuconolam. Arkivoc **3**: 163
262. Kam TS, Tee YM, Subramaniam G (1998) Rhazinal, a Formylrhazinilam Derivative from a Malayan *Kopsia*. Nat Prod Lett **12**: 307
263. Bowie AL, Hughes CC, Trauner D (2005) Concise Synthesis of (+/–)-Rhazinilam Through Direct Coupling. Org Lett **7**: 5207
264. Bowie AL, Trauner D (2009) Total Synthesis of (+/–)-Rhazinal Using Novel Palladium-Catalyzed Cyclizations. J Org Chem **74**: 1581
265. David B, Sevenet T, Thoison O, Awang K, Pais M, Wright M, Guenard D (1997) Hemisynthesis of Rhazinilam Analogues: Structure-Activity Relationships on Tubulin-Microtubule System. Bioorg Med Chem Lett **7**: 2155

266. Baudoin O, Claveau F, Thoret S, Herrbach A, Guenard D, Gueritte F (2002) Synthesis and Biological Evaluation of A-Ring Biaryl-Carbamate Analogues of Rhazinilam. Bioorg Med Chem **10**: 3395

267. Kim S-G, Kim J, Jung H (2005) Efficient Total Synthesis of (+)-Curcuphenol *via* Asymmetric Organocatalysis. Tetrahedron Lett **46**: 2437

268. Hong L, Wang L, Sun WS, Wong KY, Wang R (2009) Organocatalytic Asymmetric *Friedel-Crafts* Alkylation/Cyclization Cascade Reaction of 1-Naphthols and α,β-Unsaturated Aldehydes: An Enantioselective Synthesis of Chromanes and Dihydrobenzopyranes. J Org Chem **74**: 6881

269. Wright AE, Pomponi SA, McConnell OJ, Kohmoto S, McCarthy PJ (1987) (+)-Curcuphenol and (+)-Curcudiol, Sesquiterpene Phenols from Shallow and Deep-Water Collections of the Marine Sponge *Didiscus flavus*. J Nat Prod **50**: 976

270. Fusetani N, Sugano M, Matsunaga S, Hashimoto K (1987) (+)-Curcuphenol and Dehydrocurcuphenol, Novel Sesquiterpenes Which Inhibit H, K-ATPase, from a Marine Sponge *Epipolasis* sp. Experientia **43**: 1234

271. Knowles RR, Carpenter J, Blakey SB, Kayano A, Mangion IK, Sinz CJ, MacMillan DWC (2011) Total Synthesis of Diazonamide A. Chem Sci **2**: 308

272. Lindquist N, Fenical W, Vanduyne GD, Clardy J (1991) Isolation and Structure Determination of Diazonamide A and Diazonamide B, Unusual Cytotoxic Metabolites from the Marine Ascidian *Diazona chinensis*. J Am Chem Soc **113**: 2303

273. Fernandez R, Martin MJ, Rodriguez-Acebes R, Reyes F, Francesch A, Cuevas C (2008) Diazonamides C-E, New Cytotoxic Metabolites from the Ascidian *Diazona* sp. Tetrahedron Lett **49**: 2283

274. Cruz-Monserrate Z, Vervoort HC, Bai RL, Newman DJ, Howell SB, Los G, Mullaney JT, Williams MD, Pettit GR, Fenical W, Hamel E (2003) Diazonamide A and a Synthetic Structural Analog: Disruptive Effects on Mitosis and Cellular Microtubules and Analysis of their Interactions with Tubulin. Mol Pharmacol **63**: 1273

275. Nicolaou KC, Rao PB, Hao JL, Reddy MV, Rassias G, Huang XH, Chen DYK, Snyder SA (2003) The Second Total Synthesis of Diazonamide A. Angew Chem Int Ed **42**: 1753

276. Burgett AWG, Li QY, Wei Q, Harran PG (2003) A Concise and Flexible Total Synthesis of (–)-Diazonamide A. Angew Chem Int Ed **42**: 4961

277. Nicolaou KC, Hao JL, Reddy MV, Rao PB, Rassias G, Snyder SA, Huang XH, Chen DYK, Brenzovich WE, Giuseppone N, Giannakakou P, O'Brate A (2004) Chemistry and Biology of Diazonamide A: Second Total Synthesis and Biological Investigations. J Am Chem Soc **126**: 12897

278. Lee S, MacMillan DWC (2007) Organocatalytic Vinyl and *Friedel-Crafts* Alkylations with Trifluoroborate Salts. J Am Chem Soc **129**: 15438

279. Reiter M, Torssell S, Lee S, MacMillan DWC (2010) The Organocatalytic Three-Step Total Synthesis of (+)-Frondosin B. Chem Sci **1**: 37

280. Patil AD, Freyer AJ, Killmer L, Offen P, Carte B, Jurewicz AJ, Johnson RK (1997) Frondosins, Five New Sesquiterpene Hydroquinone Derivatives with Novel Skeletons from the Sponge *Dysidea frondosa*: Inhibitors of Interleukin-8 Receptors. Tetrahedron **53**: 5047

281. Hallock YF, Cardellina JH, II, Boyd MR (1998) (–)-Frondosins A and D, HIV-Inhibitory Sesquiterpene Hydroquinone Derivatives from *Euryspongia* sp. Nat Prod Lett **11**: 153

282. Inoue M, Carson MW, Frontier AJ, Danishefsky SJ (2001) Total Synthesis and Determination of the Absolute Configuration of Frondosin B. J Am Chem Soc **123**: 1878

283. Hughes CC, Trauner D (2002) Concise Total Synthesis of (–)-Frondosin B Using a Novel Palladium-Catalyzed Cyclization. Angew Chem Int Ed **41**: 1569

284. Kerr DJ, Willis AC, Flynn BL (2004) Multicomponent Coupling Approach to (+/–)-Frondosin B and a Ring-Expanded Analogue. Org Lett **6**: 457

285. Ovaska TV, Sullivan JA, Ovaska SI, Winegrad JB, Fair JD (2009) Asymmetric Synthesis of Seven-Membered Carbocyclic Rings *via* a Sequential Oxyanionic 5-*Exo*-Dig Cyclization/*Claisen* Rearrangement Process. Total Synthesis of (–)-Frondosin B. Org Lett **11**: 2715

286. Enders D, Wang C, Liebich JX (2009) Organocatalytic Asymmetric Aza-*Michael* Additions. Chem Eur J **15**: 11058

287. Krishna PR, Sreeshailam A, Srinivas R (2009) Recent Advances and Applications in Asymmetric Aza-*Michael* Addition Chemistry. Tetrahedron **65**: 9657

288. Fustero S, Jimenez D, Moscardo J, Catalan S, del Pozo C (2007) Enantioselective Organocatalytic Intramolecular Aza-*Michael* Reaction: a Concise Synthesis of (+)-Sedamine, (+)-Allosedamine, and (+)-Coniine. Org Lett **9**: 5283

289. Fustero S, Moscardo J, Jimenez D, Perez-Carrion MD, Sanchez-Rosello M, del Pozo C (2008) Organocatalytic Approach to Benzofused Nitrogen-Containing Heterocycles: Enantioselective Total Synthesis of (+)-Angustureine. Chem Eur J **14**: 9868

290. Reynolds T (2005) Hemlock Alkaloids from *Socrates* to Poison Aloes. Phytochemistry **66**: 1399

291. Jacquemond-Collet I, Hannedouche S, Fabre N, Fouraste I, Moulis C (1999) Two Tetrahydroquinoline Alkaloids from *Galipea officinalis*. Phytochemistry **51**: 1167

292. Marigo M, Franzen J, Poulsen TB, Zhuang W, Jørgensen KA (2005) Asymmetric Organocatalytic Epoxidation of α,β-Unsaturated Aldehydes with Hydrogen Peroxide. J Am Chem Soc **127**: 6964

293. Nicolaou KC, Sarlah D, Wu TR, Zhan WQ (2009) Total Synthesis of Hirsutellone B. Angew Chem Int Ed **48**: 6870

294. Isaka M, Rugseree N, Maithip P, Kongsaeree P, Prabpai S, Thebtaranonth Y (2005) Hirsutellones A-E, Antimycobacterial Alkaloids from the Insect Pathogenic Fungus *Hirsutella nivea* BCC 2594. Tetrahedron **61**: 5577

295. Isaka M, Prathumpai W, Wongsa P, Tanticharoen M (2006) Hirsutellone F, a Dimer of Antitubercular Alkaloids from the Seed Fungus *Trichoderma* Species BCC 7579. Org Lett **8**: 2815

296. Bertelsen S, Diner P, Johansen RL, Jørgensen KA (2007) Asymmetric Organocatalytic β-Hydroxylation of α,β-Unsaturated Aldehydes. J Am Chem Soc **129**: 1536

297. Andersen NR, Hansen SG, Bertelsen S, Jørgensen KA (2009) Organocatalysis in Natural Product Synthesis: A Simple One-Pot Approach to Optically Active β-Diols. Adv Synth Catal **351**: 3193

298. Tan KH (2000) Sex Pheromone Components in Defense of Melon Fly, *Bactrocera cucurbitae* Against Asian House Gecko, *Hemidactylus frenatus*. J Chem Ecol **26**: 697

299. Jones SB, Simmons B, Mastracchio A, MacMillan DWC (2011) Collective Synthesis of Natural Products by Means of Organocascade Catalysis. Nature **475**: 183

300. Hudlicky T, Reed JW (2007) The Way of Synthesis: Evolution of Design and Methods for Natural Products. Wiley-VCH, Weinheim

301. Hong BC, Wu MF, Tseng HC, Liao JH (2006) Enantioselective Organocatalytic Formal [3+3]-Cycloaddition of α,β-Unsaturated Aldehydes and Application to the Asymmetric Synthesis of (–)-Isopulegol Hydrate and (–)-Cubebaol. Org Lett **8**: 2217

302. Hong BC, Nimje RY, Yang CY (2007) The Organocatalytic Direct Self-Trimerization of Acrolein: Application to the Total Synthesis of Montiporyne F. Tetrahedron Lett **48**: 1121

303. Michrowska A, List B (2009) Concise Synthesis of Ricciocarpin A and Discovery of a More Potent Analogue. Nat Chem **1**: 225

304. Simmons B, Walji AM, MacMillan DWC (2009) Cycle-Specific Organocascade Catalysis: Application to Olefin Hydroamination, Hydro-oxidation, and Amino-oxidation, and to Natural Product Synthesis. Angew Chem Int Ed **48**: 4349

305. Hong BC, Kotame P, Tsai CW, Liao JH (2010) Enantioselective Total Synthesis of (+)-Conicol *via* Cascade Three-Component Organocatalysis. Org Lett **12**: 776

306. Vanek T, Novotny M, Podlipna R, Saman D, Valterova I (2003) Biotransformation of Citronellal by *Solanum aviculare* Suspension Cultures: Preparation of *p*-Menthane-3,8-diols and Determination of their Absolute Configurations. J Nat Prod **66**: 1239

307. Bae BH, Im KS, Choi WC, Hong JK, Lee CO, Choi JS, Son BW, Song JI, Jung JH (2000) New Acetylenic Compounds from the Stony Coral *Montipora* sp. J Nat Prod **63**: 1511

308. Wurzel G, Becker H (1990) Sesquiterpenoids from the Liverwort *Ricciocarpos natans*. Phytochemistry **29**: 2565

309. Zinsmeister HD, Becker H, Eicher T (1991) Bryophytes, a Source of Biologically Active, Naturally Occurring Material. Angew Chem Int Ed **30**: 130

310. Garrido L, Zubia E, Ortega MJ, Salva J (2002) New Meroterpenoids From the Ascidian *Aplidium conicum*. J Nat Prod **65**: 1328

311. Simon-Levert A, Arrault A, Bontemps-Subielos N, Canal C, Banaigs B (2005) Meroterpenes From the Ascidian *Aplidium* aff. *densum*. J Nat Prod **68**: 1412

312. Walji AM, MacMillan DWC (2007) Strategies to Bypass the Taxol Problem. Enantioselective Cascade Catalysis, a New Approach for the Efficient Construction of Molecular Complexity. Synlett 1477

313. Beechan CM, Djerassi C, Eggert H (1978) Terpenoids 74. Sesquiterpenes from Soft Coral *Sinularia mayi*. Tetrahedron **34**: 2503

314. Moreira IC, Lago JHG, Young MCM, Roque NF (2003) Antifungal Aromadendrane Sesquiterpenoids from the Leaves of *Xylopia brasiliensis*. J Brazil Chem Soc **14**: 828

315. Wu TS, Chan YY, Leu YL (2000) The Constituents of the Root and Stem of *Aristolochia heterophylla* Hemsl. Chem Pharm Bull **48**: 357

316. Beeson TD, Mastracchio A, Hong JB, Ashton K, MacMillan DWC (2007) Enantioselective Organocatalysis Using SOMO Activation. Science **316**: 582

317. Amatore M, Beeson TD, Brown SP, MacMillan DWC (2009) Enantioselective Linchpin Catalysis by SOMO Catalysis: An Approach to the Asymmetric α-Chlorination of Aldehydes and Terminal Epoxide Formation. Angew Chem Int Ed **48**: 5121

318. Um JM, Gutierrez O, Schoenebeck F, Houk KN, MacMillan DWC (2010) Nature of Intermediates in Organo-SOMO Catalysis of α-Arylation of Aldehydes. J Am Chem Soc **132**: 6001

319. Mastracchio A, Warkentin AA, Walji AM, MacMillan DWC (2010) Direct and Enantioselective α-Allylation of Ketones *via* Singly Occupied Molecular Orbital (SOMO) Catalysis. Proc Natl Acad Sci USA **107**: 20648

320. Jui NT, Lee ECY, MacMillan DWC (2010) Enantioselective Organo-SOMO Cascade Cycloadditions: A Rapid Approach to Molecular Complexity from Simple Aldehydes and Olefins. J Am Chem Soc **132**: 10015

321. Nicolaou KC, Reingruber R, Sarlah D, Bräse S (2009) Enantioselective Intramolecular *Friedel-Crafts*-Type α-Arylation of Aldehydes. J Am Chem Soc **131**: 2086

322. Winter P, Vaxelaire C, Heinz C, Christmann M (2011) Transforming Terpene Feedstock into Polyketide Architecture. Chem Commun **47**: 394

323. Bohlmann F, Zdero C, Robinson H, King RM (1979) Natural Terpene Derivatives. 217. New Cadinene and Norcadinene Derivatives from *Heterotheca grandiflora*. Phytochemistry **18**: 1675

324. Irschik H, Augustiniak H, Gerth K, Höfle G, Reichenbach H (1995) The Ripostatins, Novel Inhibitors of Eubacterial RNA-Polymerase Isolated from Myxobacteria. J Antibiot **48**: 787

325. Starks CM (1971) Phase-Transfer Catalysis. 1. Heterogeneous Reactions Involving Anion Transfer by Quaternary Ammonium and Phosphonium Salts. J Am Chem Soc **93**: 195

326. Dehmlow EW, Dehmlow SS (1993) Phase Transfer Catalysis, 3rd edn. VCH, Weinheim

327. Starks CM, Liotta CL, Halpern ME (1994) Phase-Transfer Catalysis. Chapman and Hall, New York

328. O'Donnell MJ (2000) Asymmetric Phase-Transfer Reactions. In: Ojima I (ed) Catalytic Asymmetric Syntheses, 2nd edn. Wiley-VCH, New York, p 727

329. Maruoka K (2008) Asymmetric Phase Transfer Catalysis. Wiley-VCH, Weinheim

330. Maruoka K, Hashimoto T (2007) Recent Development and Application of Chiral Phase-Transfer Catalysts. Chem Rev **107**: 5656

331. O'Donnell MJ (2004) The Enantioselective Synthesis of α-Amino Acids by Phase-Transfer Catalysis with Achiral *Schiff* Base Esters. Acc Chem Res **37**: 506

332. Maruoka K, Ooi T (2007) Recent Advances in Asymmetric Phase-Transfer Catalysis. Angew Chem Int Ed **46**: 4222

333. Helder R, Hummelen JC, Laane RWPM, Wiering JS, Wynberg H (1976) Catalytic Asymmetric Induction in Oxidation Reactions – Synthesis of Optically Active Epoxides. Tetrahedron Lett **17**: 1831

334. O'Donnell MJ, Bennett WD, Wu SD (1989) The Stereoselective Synthesis of α-Amino Acids by Phase-Transfer Catalysis. J Am Chem Soc **111**: 2353

335. Corey EJ, Xu F, Noe MC (1997) A Rational Approach to Catalytic Enantioselective Enolate Alkylation Using a Structurally Rigidified and Defined Chiral Quaternary Ammonium Salt Under Phase Transfer Conditions. J Am Chem Soc **119**: 12414

336. Maruoka K, Ooi T, Kameda M (1999) Molecular Design of a C_2-Symmetric Chiral Phase-Transfer Catalyst for Practical Asymmetric Synthesis of α-Amino Acids. J Am Chem Soc **121**: 6519

337. Shibasaki M, Ohshima T, Shibuguchi T, Fukuta Y (2004) Catalytic Asymmetric Phase-Transfer Reactions Using Tartrate-Derived Asymmetric Two-Center Organocatalysts. Tetrahedron **60**: 7743

338. Denmark SE, Gould ND, Wolf LM (2011) A Systematic Investigation of Quaternary Ammonium Ions as Asymmetric Phase-Transfer Catalysts. Synthesis of Catalyst Libraries and Evaluation of Catalyst Activity. J Org Chem **76**: 4260

339. Denmark SE, Gould ND, Wolf LM (2011) A Systematic Investigation of Quaternary Ammonium Ions as Asymmetric Phase-Transfer Catalysts. Application of Quantitative Structure Activity/Selectivity Relationships. J Org Chem **76**: 4337

340. Waser M, Gratzer K, Herchl R, Müller N (2012) Design, Synthesis, and Application of Tartaric Acid Derived *N-Spiro* Quaternary Ammonium Salts as Chiral Phase-Transfer Catalysts. Org Biomol Chem **10**: 251

341. Lee TBK, Wong GSK (1991) Asymmetric Alkylation of Oxindoles: An Approach to the Total Synthesis of (–)-Physostigmine. J Org Chem **56**: 872

342. Andrus MB, Hicken EJ, Stephens JC, Bedke DK (2006) Total Synthesis of the Hydroxyketone Kurasoin A Using Asymmetric Phase-Transfer Alkylation. J Org Chem **71**: 8651

343. Andrus MB, Christiansen MA, Butler AW, Hill AR (2009) Synthesis of Kurasoin B Using Phase-Transfer-Catalyzed Acylimidazole Alkylation. Synlett **2009**: 653

344. Uchida R, Shiomi K, Inokoshi J, Masuma R, Kawakubo T, Tanaka H, Iwai Y, Omura S (1996) Kurasoins A and B, New Protein Farnesyltransferase Inhibitors Produced by *Paecilomyces* sp. FO-3684. 1. Producing Strain, Fermentation, Isolation, and Biological Activities. J Antibiot **49**: 932

345. Quinoa E, Adamczeski M, Crews P, Bakus GJ (1986) Bengamides, Heterocyclic Anthelmintics from a Jaspidae Marine Sponge. J Org Chem **51**: 4494

346. Boeckman RK, Clark TJ, Shook BC (2002) A Practical Enantioselective Total Synthesis of the Bengamides B, E, and Z. Org Lett **4**: 2109

347. Kim S, Lee J, Lee T, Park HG, Kim D (2003) First Asymmetric Total Synthesis of (–)-Antofine by Using an Enantioselective Catalytic Phase Transfer Alkylation. Org Lett **5**: 2703

348. Staerk D, Lykkeberg AK, Christensen J, Budnik BA, Abe F, Jaroszewski JW (2002) In Vitro Cytotoxic Activity of Phenanthroindolizidine Alkaloids from *Cynanchum vincetoxicum* and *Tylophora tanakae* Against Drug-Sensitive and Multidrug-Resistant Cancer Cells. J Nat Prod **65**: 1299

349. Murakami M, Okita Y, Matsuda H, Okino T, Yamaguchi K (1994) Aeruginosin 298-A, a Thrombin and Trypsin-Inhibitor from the Blue-Green-Alga *Microcystis aeruginosa* (Nies-298). Tetrahedron Lett **35**: 3129

350. Shibasaki M, Ohshima T, Gnanadesikan V, Shibuguchi T, Fukuta Y, Nemoto T (2003) Enantioselective Syntheses of Aeruginosin 298-A and its Analogues Using a Catalytic Asymmetric Phase-Transfer Reaction and Epoxidation. J Am Chem Soc **125**: 11206

351. Stone TW, Darlington LG (2002) Endogenous Kynurenines as Targets for Drug Discovery and Development. Nat Rev Drug Discov **1**: 609

352. Lygo B, Andrews BI (2003) Enantioselective Synthesis of Aroylalanine Derivatives. Tetrahedron Lett **44**: 4499

353. Takaishi Y, Shishido K, Wariishi N, Shibuya M, Goto K, Kido M, Takai M, Ono Y (1992) Triptoquinone A and B, Novel Interleukin-1 Inhibitors from *Tripterygium wilfordii* var. *regelii*. Tetrahedron Lett **33**: 7177

354. Shishido K, Goto K, Miyoshi S, Takaishi Y, Shibuya M (1994) Synthetic Studies on Diterpenoid Quinones with Interleukin-1 Inhibitory Activity – Total Synthesis of (±)- and (+)-Triptoquinone A. J Org Chem **59**: 406

355. Shibasaki M, Shibuguchi T, Mihara H, Kuramochi A, Sakuraba S, Ohshima T (2006) Short Synthesis of (+)-Cylindricine C by Using a Catalytic Asymmetric *Michael* Reaction with a Two-Center Organocatalyst. Angew Chem Int Ed **45**: 4635

356. Shibuguchi T, Mihara H, Kuramochi A, Ohshima T, Shibasaki M (2007) Catalytic Asymmetric Phase-Transfer *Michael* Reaction and *Mannich*-Type Reaction of Glycine *Schiff* Bases with Tartrate-Derived Diammonium Salts. Chem Asian J **2**: 794

357. Mihara H, Shibuguchi T, Kuramochi A, Ohshima T, Shibasaki M (2007) Short Synthesis of (+)-Cylindricine C and Formal Total Synthesis of (–)-Lepadiformine. Heterocycles **72**: 421

358. Blackman AJ, Li CP, Hockless DCR, Skelton BW, White AH (1993) Cylindricine A and Cylindricine B, Novel Alkaloids from the Ascidian *Clavelina cylindrica*. Tetrahedron **49**: 8645

359. Li CP, Blackman AJ (1994) Cylindricines C-G, Perhydropyrrolo[2,1-J]quinolin-7-one Alkaloids from the Ascidian *Clavelina cylindrica*. Aust J Chem **47**: 1355

360. Li CP, Blackman AJ (1995) Cylindricines H-K, Novel Alkaloids from the Ascidian *Clavelina cylindrica*. Aust J Chem **48**: 955

361. Biard JF, Guyot S, Roussakis C, Verbist JF, Vercauteren J, Weber JF, Boukef K (1994) Lepadiformine, a New Marine Cytotoxic Alkaloid from *Clavelina lepadiformis* Müller. Tetrahedron Lett **35**: 2691

362. Qi J, Beeler AB, Zhang QA, Porco JA, Jr (2010) Catalytic Enantioselective Alkylative Dearomatization-Annulation: Total Synthesis and Absolute Configuration Assignment of Hyperibone K. J Am Chem Soc **132**: 13642

363. Maciver EE, Thompson S, Smith MD (2009) Catalytic Asymmetric 6π Electrocyclization: Enantioselective Synthesis of Functionalized Indolines. Angew Chem Int Ed **48**: 9979

364. Kano T, Yamamoto A, Song S, Maruoka K (2011) Catalytic Asymmetric Syntheses of Isoxazoline-*N*-Oxides under Phase-Transfer Conditions. Chem Commun **47**: 4358

365. Tanaka N, Takaishi Y, Shikishima Y, Nakanishi Y, Bastow K, Lee KH, Honda G, Ito M, Takeda Y, Kodzhimatov OK, Ashurmetov O (2004) Prenylated Benzophenones and Xanthones from *Hypericum scabrum*. J Nat Prod **67**: 187

366. Taylor MS, Jacobsen EN (2006) Asymmetric Catalysis by Chiral Hydrogen-Bond Donors. Angew Chem Int Ed **45**: 1520

367. Doyle AG, Jacobsen EN (2007) Small-Molecule H-Bond Donors in Asymmetric Catalysis. Chem Rev **107**: 5713

368. Yu XH, Wang W (2008) Hydrogen-Bond-Mediated Asymmetric Catalysis. Chem Asian J **3**: 516

369. Akiyama T, Itoh J, Fuchibe K (2006) Recent Progress in Chiral *Brønsted* Acid Catalysis. Adv Synth Catal **348**: 999

370. Cheon CH, Yamamoto H (2011) Super *Brønsted* Acid Catalysis. Chem Commun **47**: 3043

371. Rueping M, Kuenkel A, Atodiresei I (2011) Chiral *Brønsted* Acids in Enantioselective Carbonyl Activations – Activation Modes and Applications. Chem Soc Rev **40**: 4539

372. Rueping M, Nachtsheim BJ, Ieawsuwan W, Atodiresei I (2011) Modulating the Acidity: Highly Acidic *Brønsted* Acids in Asymmetric Catalysis. Angew Chem Int Ed **50**: 6706

373. Aleman J, Parra A, Jiang H, Jørgensen KA (2011) Squaramides: Bridging from Molecular Recognition to Bifunctional Organocatalysis. Chem Eur J **17**: 6890

374. Wassermann A (1942) Homogeneous Catalysis of Diene Syntheses. A New Type of Third-Order Reaction. J Chem Soc 618

375. Yates P, Eaton P (1960) Acceleration of the *Diels-Alder* Reaction by Aluminum Chloride. J Am Chem Soc **82**: 4436
376. Rueping M, Antonchick AR, Theissmann T (2006) A Highly Enantioselective *Brønsted* Acid Catalyzed Cascade Reaction: Organocatalytic Transfer Hydrogenation of Quinolines and Their Application in the Synthesis of Alkaloids. Angew Chem Int Ed **45**: 3683
377. Rakotoson JH, Fabre N, Jacquemond-Collet I, Hannedouche S, Fouraste I, Moulis C (1998) Alkaloids from *Galipea officinalis*. Planta Med **64**: 762
378. Houghton PJ, Woldemariam TZ, Watanabe Y, Yates W (1999) Activity Against *Mycobacterium tuberculosis* of Alkaloid Constituents of Angostura Bark, *Galipea officinalis*. Planta Med **65**: 250
379. Rueping M, Antonchick AP (2007) Organocatalytic Enantioselective Reduction of Pyridines. Angew Chem Int Ed **46**: 4562
380. Sklenicka HM, Hsung RP, McLaughlin MJ, Wei LI, Gerasyuto AI, Brennessel WB (2002) Stereoselective Formal [3+3] Cycloaddition Approach to *cis*-1-Azadecalins and Synthesis of (–)-4a,8a-*diepi*-Pumiliotoxin C. Evidence for the First Highly Stereoselective 6 π-Electron Electrocyclic Ring Closures of 1-Azatrienes. J Am Chem Soc **124**: 10435
381. Spande TF, Jain P, Garraffo HM, Pannell LK, Yeh HJC, Daly JW, Fukumoto S, Imamura K, Tokuyama T, Torres JA, Snelling RR, Jones TH (1999) Occurrence and Significance of Decahydroquinolines from Dendrobatid Poison Frogs and a Myrmicine Ant: Use of ^{1}H and ^{13}C NMR in Their Conformational Analysis. J Nat Prod **62**: 5
382. Sewgobind NV, Wanner MJ, Ingemann S, de Gelder R, van Maarseveen JH, Hiemstra H (2008) Enantioselective BINOL-Phosphoric Acid Catalyzed *Pictet-Spengler* Reactions of N-Benzyltryptamine. J Org Chem **73**: 6405
383. van Maarseveen JH, Wanner MJ, Boots RNA, Eradus B, de Gelder R, Hiemstra H (2009) Organocatalytic Enantioselective Total Synthesis of (–)-Arboricine. Org Lett **11**: 2579
384. Lim KH, Komiyama K, Kam TS (2007) Arboricine and Arboricinine, Unusual Tetracyclic Indole Regioisomers from *Kopsia*. Tetrahedron Lett **48**: 1143
385. Herle B, Wanner MJ, van Maarseveen JH, Hiemstra H (2011) Total Synthesis of (+)-Yohimbine *via* an Enantioselective Organocatalytic *Pictet-Spengler* Reaction. J Org Chem **76**: 8907
386. Tamelen EEV, Shamma M, Burgstahler AW, Wolinsky J, Tamm R, Aldrich PE (1958) The Total Synthesis of Yohimbine. J Am Chem Soc **80**: 5006
387. Goldberg MR, Robertson D (1983) Yohimbine – a Pharmacological Probe for Study of the α_2-Adrenoreceptor. Pharmacol Rev **35**: 143
388. Zuend SJ, Jacobsen EN, Mergott DJ (2008) Catalytic Asymmetric Total Synthesis of (+)-Yohimbine. Org Lett **10**: 745
389. Casiraghi G, Battistini L, Curti C, Rassu G, Zanardi F (2011) The Vinylogous Aldol and Related Addition Reactions: Ten Years of Progress. Chem Rev **111**: 3076
390. Sickert M, Schneider C (2008) The Enantioselective, *Brønsted* Acid Catalyzed, Vinylogous *Mannich* Reaction. Angew Chem Int Ed **47**: 3631
391. Giera DS, Sickert M, Schneider C (2008) *Brønsted* Acid-Catalyzed, Enantioselective, Vinylogous *Mannich* Reaction of Vinylketene Silyl N,O-Acetals. Org Lett **10**: 4259
392. Giera DS, Sickert M, Schneider C (2009) A Straightforward Synthesis of (*S*)-Anabasine *via* the Catalytic, Enantioselective Vinylogous *Mukaiyama-Mannich* Reaction. Synthesis 3797
393. Huang Y, Unni AK, Thadani AN, Rawal VH (2003) Hydrogen Bonding: Single Enantiomers from a Chiral-Alcohol Catalyst. Nature **424**: 146
394. Du HF, Zhao DB, Ding KL (2004) Enantioselective Catalysis of the Hetero-*Diels-Alder* Reaction Between *Brassard's* Diene and Aldehydes by Hydrogen-Bonding Activation: A One-Step Synthesis of (*S*)-(+)-Dihydrokawain. Chem Eur J **10**: 5964
395. Gerard B, Cencic R, Pelletier J, Porco JA, Jr (2007) Enantioselective Synthesis of the Complex Rocaglate (–)-Silvestrol. Angew Chem Int Ed **46**: 7831
396. Spino C, Mayes N, Desfosses H (1996) Enantioselective Synthesis of (+) and (–)-Dihydrokawain. Tetrahedron Lett **37**: 6503

397. Klohs MW, Keller F, Williams RE, Toekes MI, Cronheim GE (1959) A Chemical and Pharmacological Investigation of *Piper methysticum* Forst. J Med Pharm Chem **1**: 95
398. Hwang BY, Su BN, Chai HB, Mi QW, Kardono LBS, Afriastini JJ, Riswan S, Santarsiero BD, Mesecar AD, Wild R, Fairchild CR, Vite GD, Rose WC, Farnsworth NR, Cordell GA, Pezzuto JM, Swanson SM, Kinghorn AD (2004) Silvestrol and Episilvestrol, Potential Anticancer Rocaglate Derivatives from *Aglaia silvestris*. J Org Chem **69**: 3350
399. Hwang BY, Su BN, Chai HB, Mi QW, Kardono LBS, Afriastini JJ, Riswan S, Santarsiero BD, Mesecar AD, Wild R, Fairchild CR, Vite GD, Rose WC, Farnsworth NR, Cordell GA, Pezzuto JM, Swanson SM, Kinghorn AD (2004) Silvestrol and Episilvestrol, Potential Anticancer Rocaglate Derivatives from *Aglaia silvestris*. J Org Chem **69**: 6156
400. Mi QW, Kim S, Hwang BY, Su BN, Chai H, Arbieva ZH, Kinghorn AD, Swanson SM (2006) Silvestrol Regulates G_2/M Checkpoint Genes independent of p53 Activity. Anticancer Res **26**: 3349
401. Ebada SS, Lajkiewicz N, Porco JA Jr, Li-Weber M, Proksch P (2011) Chemistry and Biology of Rocaglamides (= Flavaglines) and Related Derivatives from *Aglaia* Species (Meliaceae). Prog Chem Org Nat Prod **94**: 1
402. Gerard B, Jones G, Porco JA, Jr (2004) A Biomimetic Approach to the Rocaglamides Employing Photogeneration of Oxidopyryliums Derived from 3-Hydroxyflavones. J Am Chem Soc **126**: 13620
403. Gerard B, Sangji S, O'Leary DJ, Porco JA (2006) Enantioselective Photocycloaddition Mediated by Chiral *Brønsted* Acids: Asymmetric Synthesis of the Rocaglamides. J Am Chem Soc **128**: 7754
404. Raheem IT, Thiara PS, Peterson EA, Jacobsen EN (2007) Enantioselective *Pictet-Spengler*-Type Cyclizations of Hydroxylactams: H-Bond Donor Catalysis by Anion Binding. J Am Chem Soc **129**: 13404
405. Kam TS, Sim KM (1998) Alkaloids from *Kopsia griffithii*. Phytochemistry **47**: 145
406. Chrzanowska M, Rozwadowska MD (2004) Asymmetric Synthesis of Isoquinoline Alkaloids. Chem Rev **104**: 3341
407. Kanemitsu T, Yamashita Y, Nagata K, Itoh T (2006) Catalytic Asymmetric Synthesis of (*R*)-(–)-Calycotomine, (*S*)-(–)-Salsolidine and (*S*)-(–)-Carnegine. Synlett 1595
408. Vachal P, Jacobsen EN (2002) Structure-Based Analysis and Optimization of a Highly Enantioselective Catalyst for the *Strecker* Reaction. J Am Chem Soc **124**: 10012
409. Takemoto Y, Miyabe H (2007) The Amino Thiourea-Catalyzed Asymmetric Nucleophilic Reactions. Chimia **61**: 269
410. Okino T, Hoashi Y, Takemoto Y (2003) Enantioselective *Michael* Reaction of Malonates to Nitroolefins Catalyzed by Bifunctional Organocatalysts. J Am Chem Soc **125**: 12672
411. Hoashi Y, Okino T, Takemoto Y (2005) Enantioselective *Michael* Addition to α,β-Unsaturated Imides Catalyzed by a Bifunctional Organocatalyst. Angew Chem Int Ed **44**: 4032
412. Okino T, Hoashi Y, Furukawa T, Xu XN, Takemoto Y (2005) Enantio- and Diastereoselective *Michael* Reaction of 1,3-Dicarbonyl Compounds to Nitroolefins Catalyzed by a Bifunctional Thiourea. J Am Chem Soc **127**: 119
413. Hoashi Y, Yabuta T, Yuan P, Miyabe H, Takemoto Y (2006) Enantioselective Tandem *Michael* Reaction to Nitroalkene Catalyzed by Bifunctional Thiourea: Total Synthesis of (–)-Epibatidine. Tetrahedron **62**: 365
414. Spande TF, Garraffo HM, Edwards MW, Yeh HJC, Pannell L, Daly JW (1992) Epibatidine – a Novel (Chloropyridyl)Azabicycloheptane with Potent Analgesic Activity from an Ecuadorian Poison Frog. J Am Chem Soc **114**: 3475
415. Badio B, Daly JW (1994) Epibatidine, a Potent Analgesic and Nicotinic Agonist. Mol Pharmacol **45**: 563
416. Bui T, Syed S, Barbas CF (2009) Thiourea-Catalyzed Highly Enantio- and Diastereoselective Additions of Oxindoles to Nitroolefins: Application to the Formal Synthesis of (+)-Physostigmine. J Am Chem Soc **131**: 8758

417. Jakubec P, Cockfield DM, Dixon DJ (2009) Total Synthesis of (–)-Nakadomarin A. J Am Chem Soc **131**: 16632
418. Farmer RL, Biddle MM, Nibbs AE, Huang XK, Bergan RC, Scheidt KA (2010) Concise Syntheses of the Abyssinones and Discovery of New Inhibitors of Prostate Cancer and MMP-2 Expression. ACS Med Chem Lett **1**: 400
419. Bassas O, Huuskonen J, Rissanen K, Koskinen AMP (2009) A Simple Organocatalytic Enantioselective Synthesis of Pregabalin. Eur J Org Chem **2009**: 1340
420. Hynes PS, Stupple PA, Dixon DJ (2008) Organocatalytic Asymmetric Total Synthesis of (*R*)-Rolipram and Formal Synthesis of (3*S*,4*R*)-Paroxetine. Org Lett **10**: 1389
421. Li DR, Murugan A, Falck JR (2008) Enantioselective, Organocatalytic Oxy-*Michael* Addition to γ/δ-hydroxy-α,β-Enones: Boronate-Amine Complexes as Chiral Hydroxide Synthons. J Am Chem Soc **130**: 46
422. Macleod JK, Schaffeler L (1995) A Short Enantioselective Synthesis of a Biologically Active Compound from *Persea americana*. J Nat Prod **58**: 1270
423. Dollt H, Hammann P, Blechert S (1999) Synthesis of (+)-(*S*)-Streptenol A and Biomimetic Synthesis of (2*R*,4*S*)- and (2*S*,4*S*)-2-(Pent-3-enyl)piperidin-4-ol. Helv Chim Acta **82**: 1111
424. Li H, Chen YG, Deng L (2011) *Cinchona* Alkaloids. In: Zhou QL (ed) Privileged Catalysts and Ligands in Asymmetric Catalysis. Wiley-VCH, Weinheim, p 361
425. Wang BM, Wu FH, Wang Y, Liu XF, Deng L (2007) Control of Diastereoselectivity in Tandem Asymmetric Reactions Generating Nonadjacent Stereocenters with Bifunctional Catalysis by *Cinchona* Alkaloids. J Am Chem Soc **129**: 768
426. Wang Y, Liu XF, Deng L (2006) Dual-Function *Cinchona* Alkaloid Catalysis: Catalytic Asymmetric Tandem Conjugate Addition-Protonation for the Direct Creation of Nonadjacent Stereocenters. J Am Chem Soc **128**: 3928
427. Kobayashi J, Kanda F, Ishibashi M, Shigemori H (1991) Manzacidins A-C, Novel Tetrahydropyrimidine Alkaloids from the Okinawan Marine Sponge *Hymeniacidon* sp. J Org Chem **56**: 4574
428. Faulkner DJ (1998) Marine Natural Products. Nat Prod Rep **15**: 113
429. Hashimoto T, Maruoka K (2008) Syntheses of Manzacidins: a Stage for the Demonstration of Synthetic Methodologies. Org Biomol Chem **6**: 829
430. Tran K, Lombardi PJ, Leighton JL (2008) An Efficient Asymmetric Synthesis of Manzacidin C. Org Lett **10**: 3165
431. Kobayashi J, Watanabe D, Kawasaki N, Tsuda M (1997) Nakadomarin A, a Novel Hexacyclic Manzamine-Related Alkaloid from *Amphimedon* Sponge. J Org Chem **62**: 9236
432. Kobayashi J, Tsuda M, Ishibashi M (1999) Bioactive Products from Marine Micro- and Macro-organisms. Pure Appl Chem **71**: 1123
433. Jakubec P, Kyle AF, Calleja J, Dixon DJ (2011) Total Synthesis of (–)-Nakadomarin A: Alkyne Ring-Closing Metathesis. Tetrahedron Lett **52**: 6094
434. Kamat VS, Chuo FY, Kubo I, Nakanishi K (1981) Anti-microbial Agents from an East African Medicinal Plant *Erythrina abyssinica*. Heterocycles **15**: 1163
435. Lee D, Bhat KPL, Fong HHS, Farnsworth NR, Pezzuto JM, Kinghorn AD (2001) Aromatase Inhibitors from *Broussonetia papyrifera*. J Nat Prod **64**: 1286
436. Chen P, Bao X, Zhang LF, Ding M, Han XJ, Li J, Zhang GB, Tu YQ, Fan CA (2011) Asymmetric Synthesis of Bioactive Hydrodibenzofuran Alkaloids: (–)-Lycoramine, (–)-Galanthamine, and (+)-Lunarine. Angew Chem Int Ed **50**: 8161
437. Kondo H, Tomimura K, Ishiwata S (1932) Alkaloids of *Lycoris radiata* Herb. V and VI. Yakugaku Zasshi **52**: 433
438. Proskurnina NF, Yakovleva AP (1952) Alkaloids of *Galanthus woronowi*. II. Isolation of a New Alkaloid. Zh Obshch Khim **22**: 1899
439. Marco-Contelles J, Carreiras MD, Rodriguez C, Villarroya M, Garcia AG (2006) Synthesis and Pharmacology of Galantamine. Chem Rev **106**: 116
440. Han SY, Sweeney JE, Bachman ES, Schweiger EJ, Forloni G, Coyle JT, Davis BM, Joullie MM (1992) Chemical and Pharmacological Characterization of Galanthamine, an

Acetylcholinesterase Inhibitor, and Its Derivatives – a Potential Application in *Alzheimer's* Disease. Eur J Med Chem **27**: 673

441. Reeb E (1911) *Lunaria annua* and its Active Principle. Les Nouv Remedes **27**: 481
442. Hamilton CJ, Saravanamuthu A, Poupat C, Fairlamb AH, Eggleston IM (2006) Time Dependent Inhibitors of Trypanothione Reductase: Analogues of the Spermidine Alkaloid Lunarine and Related Natural Products. Bioorg Med Chem **14**: 2266
443. Tian SK, Chen YG, Hang JF, Tang L, McDaid P, Deng L (2004) Asymmetric Organic Catalysis with Modified *Cinchona* Alkaloids. Acc Chem Res **37**: 621
444. Fu GC (2004) Asymmetric Catalysis with "Planar-Chiral" Derivatives of 4-(Dimethylamino) pyridine. Acc Chem Res **37**: 542
445. Marcelli T, van Maarseveen JH, Hiemstra H (2006) Cupreines and Cupreidines: An Emerging Class of Bifunctional *Cinchona* Organocatalysts. Angew Chem Int Ed **45**: 7496
446. France S, Guerin DJ, Miller SJ, Lectka T (2003) Nucleophilic Chiral Amines as Catalysts in Asymmetric Synthesis. Chem Rev **103**: 2985
447. Enders D, Niemeier O, Henseler A (2007) Organocatalysis by N-Heterocyclic Carbenes. Chem Rev **107**: 5606
448. Denmark SE, Beutner GL (2008) *Lewis* Base Catalysis in Organic Synthesis. Angew Chem Int Ed **47**: 1560
449. Waldmann H, Khedkar V, Duckert H, Schumann M, Oppel IM, Kumar K (2008) Asymmetric Synthesis of Natural Product Inspired Tricyclic Benzopyrones by an Organocatalyzed Annulation Reaction. Angew Chem Int Ed **47**: 6869
450. Hiemstra H, Wynberg H (1981) Addition of Aromatic Thiols to Conjugated Cycloalkenones, Catalyzed by Chiral β-Hydroxy Amines – a Mechanistic Study on Homogeneous Catalytic Asymmetric Synthesis. J Am Chem Soc **103**: 417
451. Sekino E, Kumamoto T, Tanaka T, Ikeda T, Ishikawa T (2004) Concise Synthesis of Anti-HIV-1 Active (+)-Inophyllum B and (+)-Calanolide A by Application of (–)-Quinine-Catalyzed Intramolecular Oxo-*Michael* Addition. J Org Chem **69**: 2760
452. Kashman Y, Gustafson KR, Fuller RW, Cardellina JH, II, McMahon JB, Currens MJ, Buckheit RW, Hughes SH, Cragg GM, Boyd MR (1992) The Calanolides, a Novel HIV-inhibitory Class of Coumarin Derivatives from the Tropical Rainforest Tree, *Calophyllum lanigerum*. J Med Chem **35**: 2735
453. Patil AD, Freyer AJ, Eggleston DS, Haltiwanger RC, Bean MF, Taylor PB, Caranfa MJ, Breen AL, Bartus HR, Johnson RK, Hertzberg RP, Westley JW (1993) The Inophyllums, Novel Inhibitors of HIV-1 Reverse-Transcriptase Isolated from the Malaysian Tree, *Calophyllum inophyllum* Linn. J Med Chem **36**: 4131
454. de Figueiredo RM, Frohlich R, Christmann M (2007) Efficient Synthesis and Resolution of Pyrrolizidines. Angew Chem Int Ed **46**: 2883
455. Lambert TH, Danishefsky SJ (2006) Total Synthesis of UCS1025A. J Am Chem Soc **128**: 426
456. Nakai R, Ogawa H, Asai A, Ando K, Agatsuma T, Matsumiya S, Akinaga S, Yamashita Y, Mizukami T (2000) UCS1025A, a Novel Antibiotic Produced by *Acremonium* sp. J Antibiot **53**: 294
457. Agatsuma T, Akama T, Nara S, Matsumiya S, Nakai R, Ogawa H, Otaki S, Ikeda S, Saitoh Y, Kanda Y (2002) UCS1025A and B, New Antitumor Antibiotics from the Fungus *Acremonium* sp. Org Lett **4**: 4387
458. Liu L, Zhang SL, Xue F, Lou GS, Zhang HY, Ma SC, Duan WH, Wang W (2011) Catalytic Enantioselective *Henry* Reactions of Isatins: Application in the Concise Synthesis of (*S*)-(–)-Spirobrassinin. Chem Eur J **17**: 7791
459. Takasugi M, Monde K, Katsui N, Shirata A (1987) Spirobrassinin, a Novel Sulfur-Containing Phytoalexin from the Daikon *Rhaphanus sativus* L. var. *hortensis* (Cruciferae). Chem Lett 1631
460. Pedras MSC, Okanga FI, Zaharia IL, Khan AQ (2000) Phytoalexins from Crucifers: Synthesis, Biosynthesis, and Biotransformation. Phytochemistry **53**: 161

461. Pedras MSC, Hossain M (2006) Metabolism of Crucifer Phytoalexins in *Sclerotinia sclerotiorum*: Detoxification of Strongly Antifungal Compounds Involves Glucosylation. Org Biomol Chem **4**: 2581

462. Mehta RG, Liu JF, Constantinou A, Hawthorne M, Pezzuto JM, Moon RC, Moriarty RM (1994) Structure-Activity-Relationships of Brassinin in Preventing the Development of Carcinogen-Induced Mammary Lesions in Organ Culture. Anticancer Res **14**: 1209

463. Calter MA, Bi FC (2000) Catalytic, Asymmetric Synthesis of the C1′-C10′ Segment of Pamamycin 621A. Org Lett **2**: 1529

464. Calter MA, Liao WS, Struss JA (2001) Catalytic, Asymmetric Synthesis of Siphonarienal. J Org Chem **66**: 7500

465. Calter MA, Guo X (2002) Synthesis of the C21-C34-Segment of the Aplyronines Using the Dimer of Methylketene. Tetrahedron **58**: 7093

466. Calter MA, Liao WS (2002) First Total Synthesis of a Natural Product Containing a Chiral, β-Diketone: Synthesis and Stereochemical Reassignment of Siphonarienedione and Siphonarienolone. J Am Chem Soc **124**: 13127

467. Calter MA (1996) Catalytic, Asymmetric Dimerization of Methylketene. J Org Chem **61**: 8006

468. Papageorgiou CD, de Dios MAC, Ley SV, Gaunt MJ (2004) Enantioselective Organocatalytic Cyclopropanation *via* Ammonium Ylides. Angew Chem Int Ed **43**: 4641

469. Bremeyer N, Smith SC, Ley SV, Gaunt MJ (2004) An intramolecular Organocatalytic Cyclopropanation Reaction. Angew Chem Int Ed **43**: 2681

470. Johansson CCC, Bremeyer N, Ley SV, Owen DR, Smith SC, Gaunt MJ (2006) Enantioselective Catalytic Intramolecular Cyclopropanation Using Modified *Cinchona* Alkaloid Organocatalysts. Angew Chem Int Ed **45**: 6024

471. Waser M, Herchl R, Müller N (2011) Ammonium Ylides for the Diastereoselective Synthesis of Glycidic Amides. Chem Commun **47**: 2170

472. Herchl R, Stiftinger M, Waser M (2011) Identification of the Best-Suited Leaving Group for the Diastereoselective Synthesis of Glycidic Amides from Stabilised Ammonium Ylides and Aldehydes. Org Biomol Chem **9**: 7023

473. Kumaraswamy G, Padmaja M (2008) Enantioselective Total Synthesis of Eicosanoid and its Congener, Using Organocatalytic Cyclopropanation, and Catalytic Asymmetric Transfer Hydrogenation Reactions as Key Steps. J Org Chem **73**: 5198

474. Kumaraswamy G, Ramakrishna G, Sridhar B (2011) Enantioselective Synthesis of Cyclopropyl δ-Lactonealdehydes and Dodecyl-5-ene-1-yne-3-ol: Advanced Intermediates of Solandelactone A and B. Tetrahedron Lett **52**: 1778

475. Gerwick WH (1993) Carbocyclic Oxylipins of Marine Origin. Chem Rev **93**: 1807

476. Seo YW, Cho KW, Rho JR, Shin JH, Kwon BM, Bok SH, Song JI (1996) Solandelactones A-I, Lactonized Cyclopropyl Oxylipins Isolated from the Hydroid *Solanderia secunda*. Tetrahedron **52**: 10583

477. Masson G, Housseman C, Zhu JP (2007) The Enantioselective *Morita-Baylis-Hillman* Reaction and its Aza Counterpart. Angew Chem Int Ed **46**: 4614

478. Ma GN, Jiang JJ, Shi M, Wei Y (2009) Recent Extensions of the *Morita-Baylis-Hillman* Reaction. Chem Commun **2009**: 5496

479. Iwabuchi Y, Sugihara T, Esumi T, Hatakeyama S (2001) An Enantio- and Stereocontrolled Route to Epopromycin B *via Cinchona* Alkaloid-Catalyzed *Baylis-Hillman* Reaction. Tetrahedron Lett **42**: 7867

480. Tsuchiya K, Kobayashi S, Nishikiori T, Nakagawa T, Tatsuta K (1997) Epopromycins, Novel Cell Wall Synthesis Inhibitors of Plant Protoplast Produced by *Streptomyces* sp. NK04000. J Antibiot **50**: 261

481. Iwabuchi Y, Furukawa M, Esumi T, Hatakeyama S (2001) An Enantio- and Stereocontrolled Synthesis of (–)-Mycestericin E *via Cinchona* Alkaloid-Catalyzed Asymmetric *Baylis-Hillman* Reaction. Chem Commun 2030

482. Sarkar SM, Wanzala EN, Shibahara S, Takahashi K, Ishihara J, Hatakeyama S (2009) Enantio- and Stereoselective Route to the Phoslactomycin Family of Antibiotics: Formal Synthesis of (+)-Fostriecin and (+)-Phoslactomycin B. Chem Commun 5907

483. Shibasaki M, Kanai M (2005) Synthetic Strategies of Fostriecin. Heterocycles **66**: 727

484. Druais V, Hall MJ, Corsi C, Wendeborn SV, Meyer C, Cossy J (2009) A Convergent Approach Toward the C1-C11 Subunit of Phoslactomycins and Formal Synthesis of Phoslactomycin B. Org Lett **11**: 935

485. Wang YG, Takeyama R, Kohayashi Y (2006) Total Synthesis of Phoslactomycin B and its Biosynthetic Deamino Precursor. Angew Chem Int Ed **45**: 3320

486. Choi CH, Tian SK, Deng L (2001) A Formal Catalytic Asymmetric Synthesis of (+)-Biotin with Modified *Cinchona* Alkaloids. Synthesis 1737

487. Gerfaud T, Xie CS, Neuville L, Zhu JP (2011) Protecting-Group-Free Total Synthesis of (*E*)- and (*Z*)-Alstoscholarine. Angew Chem Int Ed **50**: 3954

488. Cai XH, Du ZZ, Luo XD (2007) Unique Monoterpenoid Indole Alkaloids from *Alstonia scholaris*. Org Lett **9**: 1817

489. Kaneko S, Yoshino T, Katoh T, Terashima S (1998) Synthetic Studies of Huperzine A and its Fluorinated Analogues. 1. Novel Asymmetric Syntheses of an Enantiomeric Pair of Huperzine A. Tetrahedron **54**: 5471

490. Wu FH, Hong R, Khan JH, Liu XF, Deng L (2006) Asymmetric Synthesis of Chiral Aldehydes by Conjugate Additions with Bifunctional Organocatalysis by *Cinchona* Alkaloids. Angew Chem Int Ed **45**: 4301

491. Singh IP, Milligan KE, Gerwick WH (1999) Tanikolide, a Toxic and Antifungal Lactone from the Marine Cyanobacterium *Lyngbya majuscula*. J Nat Prod **62**: 1333

492. Tian SK, Hong R, Deng L (2003) Catalytic Asymmetric Cyanosilylation of Ketones with Chiral *Lewis* Base. J Am Chem Soc **125**: 9900

493. Nicolaou KC, Simonsen KB, Vassilikogiannakis G, Baran PS, Vidali VP, Pitsinos EN, Couladouros EA (1999) Biomimetic Explorations Towards the Bisorbicillinoids: Total Synthesis of Bisorbicillinol, Bisorbibutenolide, and Trichodimerol. Angew Chem Int Ed **38**: 3555

494. Lu XY, Zhang CM, Xu ZR (2001) Reactions of Electron-Deficient Alkynes and Allenes Under Phosphine Catalysis. Acc Chem Res **34**: 535

495. Methot JL, Roush WR (2004) Nucleophilic Phosphine Organocatalysis. Adv Synth Catal **346**: 1035

496. Ye LW, Zhou J, Tang Y (2008) Phosphine-Triggered Synthesis of Functionalized Cyclic Compounds. Chem Soc Rev **37**: 1140

497. Marinetti A, Voituriez A (2010) Enantioselective Phosphine Organocatalysis. Synlett 174

498. Fujiwara Y, Fu GC (2011) Application of a New Chiral Phosphepine to the Catalytic Asymmetric Synthesis of Highly Functionalized Cyclopentenes That Bear an Array of Heteroatom-Substituted Quaternary Stereocenters. J Am Chem Soc **133**: 12293

499. Tan B, Candeias NR, Barbas CF (2011) Core-Structure-Motivated Design of a Phosphine-Catalyzed [3+2] Cycloaddition Reaction: Enantioselective Syntheses of Spirocyclopenteneoxindoles. J Am Chem Soc **133**: 4672

500. Sun JW, Fu GC (2010) Phosphine-Catalyzed Formation of Carbon-Sulfur Bonds: Catalytic Asymmetric Synthesis of γ-Thioesters. J Am Chem Soc **132**: 4568

501. Tran YS, Kwon O (2005) An Application of the Phosphine-Catalyzed [4+2] Annulation in Indole Alkaloid Synthesis: Formal Syntheses of (+/−)-Alstonerine and (+/−)-Macroline. Org Lett **7**: 4289

502. Agapiou K, Krische MJ (2003) Catalytic Crossed *Michael* Cycloisomerization of Thioenoates: Total Synthesis of (+/−)-Ricciocarpin A. Org Lett **5**: 1737

503. Wang JC, Krische MJ (2003) Intramolecular Organocatalytic [3+2] Dipolar Cycloaddition: Stereospecific Cycloaddition and the Total Synthesis of (+/−)-Hirsutene. Angew Chem Int Ed **42**: 5855

504. Du YS, Lu XY (2003) A Phosphine-Catalyzed [3+2] Cycloaddition Strategy Leading to the First Total Synthesis of (–)-Hinesol. J Org Chem **68**: 6463

505. Mergott DJ, Frank SA, Roush WR (2004) Total Synthesis of (–)-Spinosyn A. Proc Natl Acad Sci USA **101**: 11955

506. Jones RA, Krische MJ (2009) Asymmetric Total Synthesis of the Iridoid β-Glucoside (+)-Geniposide *via* Phosphine Organocatalysis. Org Lett **11**: 1849

507. Morita M, Nakanishi H, Morita H, Mihashi S, Itokawa H (1996) Structures and Spasmolytic Activities of Derivatives from Sesquiterpenes of *Alpinia speciosa* and *Alpinia japonica*. Chem Pharm Bull **44**: 1603

508. Satoh K, Nagai F, Kano I (2000) Inhibition of H+, K+-ATPase by Hinesol, a Major Component of So-jutsu, by Interaction with Enzyme in the E_1-State. Biochem Pharmacol **59**: 881

509. Inouye H, Saito S, Taguchi H, Endo T (1969) Two New Iridoglucosides Gardenoside and Geniposide from *Gardenia jasminoides*. Tetrahedron Lett **10**: 2347

510. Ueda S, Iwahashi Y, Tokuda H (1991) Production of Anti-tumor-Promoting Iridoid Glucosides in *Genipa americana* and Its Cell Cultures. J Nat Prod **54**: 1677

511. Lee MJ, Hsu JD, Wang CJ (1995) Inhibition of 12-*O*-Tetradecanoylphorbol-13-acetate-Caused Tumor Promotion in Benzo[a]pyrene-Initiated CD-1 Mouse Skin by Geniposide. Anticancer Res **15**: 411

512. Kirst HA, Michel KH, Martin JW, Creemer LC, Chio EH, Yao RC, Nakatsukasa WM, Boeck L, Occolowitz JL, Paschal JW, Deeter JB, Jones ND, Thompson GD (1991) A83543a-D, Unique Fermentation-Derived Tetracyclic Macrolides. Tetrahedron Lett **32**: 4839

513. Enders D, Balensiefer T (2004) Nucleophilic Carbenes in Asymmetric Organocatalysis. Acc Chem Res **37**: 534

514. Zeitler K (2005) Extending Mechanistic Routes in Heterazolium Catalysis – Promising Concepts for Versatile Synthetic Methods. Angew Chem Int Ed **44**: 7506

515. Enders D, Niemeier O, Henseler A (2007) Organocatalysis by N-Heterocyclic Carbenes. Chem Rev **107**: 5606

516. Gillingham DG, Hoveyda AH (2007) Chiral N-Heterocyclic Carbenes in Natural Product Synthesis: Application of Ru-Catalyzed Asymmetric Ring-Opening/Cross-Metathesis and Cu-Catalyzed Allylic Alkylation to Total Synthesis of Baconipyrone C. Angew Chem Int Ed **46**: 3860

517. Harrington PE, Tius MA (2001) Synthesis and Absolute Stereochemistry of Roseophilin. J Am Chem Soc **123**: 8509

518. Christmann M (2005) New Developments in the Asymmetric *Stetter* Reaction. Angew Chem Int Ed **44**: 2632

519. Struble JR, Bode JW (2009) Formal Synthesis of Salinosporamide A *via* NHC-Catalyzed Intramolecular Lactonization. Tetrahedron **65**: 4957

520. Chiang PC, Kim Y, Bode JW (2009) Catalytic Amide Formation with α′-Hydroxyenones as Acylating Reagents. Chem Commun 4566

521. Candish L, Lupton DW (2010) The Total Synthesis of (–)-7-Deoxyloganin *via* N-Heterocyclic Carbene Catalyzed Rearrangement of α, β-Unsaturated Enol Esters. Org Lett **12**: 4836

522. Stetter H, Kuhlmann H (1975) Addition of Aldehydes to Activated Double Bonds. 7. New Simple Synthesis of *cis*-Jasmon and Dihydrojasmon. Synthesis 379

523. Trost BM, Shuey CD, Dininno F, Mcelvain SS (1979) Stereocontrolled Total Synthesis of (+/−)-Hirsutic Acid C. J Am Chem Soc **101**: 1284

524. Hayakawa Y, Kawakami K, Seto H, Furihata K (1992) Structure of a New Antibiotic, Roseophilin. Tetrahedron Lett **33**: 2701

525. Fürstner A (2003) Chemistry and Biology of Roseophilin and the Prodigiosin Alkaloids: A Survey of the Last 2500 Years. Angew Chem Int Ed **42**: 3582

526. Sharpless KB (2002) Searching for New Reactivity (*Nobel* Lecture). Angew Chem Int Ed **41**: 2024

527. Knowles WS (2002) Asymmetric Hydrogenations (*Nobel* Lecture). Angew Chem Int Ed **41**: 1999

528. Noyori R (2002) Asymmetric Catalysis: Science and Opportunities (*Nobel* Lecture). Angew Chem Int Ed **41**: 2008
529. Berkessel A (2005) Biomimetic and Organocatalytic Approaches to Oxidation Catalysis. Pure Appl Chem **77**: 1277
530. Wang ZX, Tu Y, Frohn M, Zhang JR, Shi Y (1997) An Efficient Catalytic Asymmetric Epoxidation Method. J Am Chem Soc **119**: 11224
531. Xiong ZM, Corey EJ (2000) Simple Enantioselective Total Synthesis of Glabrescol, a Chiral C_2-Symmetric Pentacyclic Oxasqualenoid. J Am Chem Soc **122**: 9328
532. Hoard DW, Moher ED, Martinelli MJ, Norman BH (2002) Synthesis of Cryptophycin 52 Using the *Shi* Epoxidation. Org Lett **4**: 1813
533. Yoshida M, Ismail MAH, Nemoto H, Ihara M (2000) Asymmetric Total Synthesis of (+)-Equilenin Utilizing Two Types of Cascade Ring Expansion Reactions of Small Ring Systems. J. Chem Soc Perkin Trans I 2629
534. Zhang QS, Lu HJ, Richard C, Curran DP (2004) Fluorous Mixture Synthesis of Stereoisomer Libraries: Total Syntheses of (+)-Murisolin and Fifteen Diastereoisomers. J Am Chem Soc **126**: 36
535. Morimoto Y, Nishikawa Y, Takaishi M (2005) Total Synthesis and Complete Assignment of the Stereostructure of a Cytotoxic Bromotriterpene Polyether (+)-Aurilol. J Am Chem Soc **127**: 5806
536. Denmark SE, Wu ZC (1999) The Development of Chiral, Nonracemic Dioxiranes for the Catalytic, Enantioselective Epoxidation of Alkenes. Synlett **1999**: 847
537. Frohn M, Shi Y (2000) Chiral Ketone-Catalyzed Asymmetric Epoxidation of Olefins. Synthesis **2000**: 1979
538. Shi Y (2004) Organocatalytic Asymmetric Epoxidation of Olefins by Chiral Ketones. Acc Chem Res **37**: 488
539. Harding WW, Lewis PA, Jacobs H, Mclean S, Reynolds WF, Tay LL, Yang JP (1995) Glabrescol – A Unique Squalene-Derived Penta-THF Diol from *Spathelia glabrescens* (Rutaceae). Tetrahedron Lett **36**: 9137
540. Xiong ZM, Corey EJ (2000) Simple Total Synthesis of the Pentacyclic C_s-Symmetric Structure Attributed to the Squalenoid Glabrescol and Three C_s-Symmetric Diastereomers Compel Structural Revision. J Am Chem Soc **122**: 4831
541. Suenaga K, Shibata T, Takada N, Kigoshi H, Yamada K (1998) Aurilol, a Cytotoxic Bromotriterpene Isolated from the Sea Hare *Dolabella auricularia*. J Nat Prod **61**: 515
542. Tong RB, Boone MA, McDonald FE (2009) Stereo- and Regioselective Synthesis of Squalene Tetraepoxide. J Org Chem **74**: 8407
543. Banfi S, Colonna S, Molinari H, Julia S, Guixer J (1984) Asymmetric Epoxidation of Electron-Poor Olefins. 5. Influence on Stereoselectivity of the Structure of Poly-α-Aminoacids Used as Catalysts. Tetrahedron **40**: 5207
544. Lasterra-Sanchez ME, Roberts SM (1997) An Important Niche in Synthetic Organic Chemistry for Some Biomimetic Oxidation Reactions Catalysed by Polyamino Acids. Curr Org Chem **1**: 187
545. Davie EAC, Mennen SM, Xu YJ, Miller SJ (2007) Asymmetric Catalysis Mediated by Synthetic Peptides. Chem Rev **107**: 5759
546. Adger BM, Barkley JV, Bergeron S, Cappi MW, Flowerdew BE, Jackson MP, McCague R, Nugent TC, Roberts SM (1997) Improved Procedure for *Julia-Colonna* Asymmetric Epoxidation of α,β-Unsaturated Ketones: Total Synthesis of Diltiazem and Taxol (TM) Side-Chain. J. Chem Soc Perkin Trans I 3501
547. Cappi MW, Chen WP, Flood RW, Liao YW, Roberts SM, Skidmore J, Smith JA, Williamson NM (1998) New Procedures for the *Julia-Colonna* Asymmetric Epoxidation: Synthesis of (+)-Clausenamide. Chem Commun 1159
548. Chen WP, Roberts SM (1999) *Julia-Colonna* Asymmetric Epoxidation of Furyl Styryl Ketone as a Route to Intermediates to Naturally Occurring Styryl Lactones. J. Chem Soc Perkin Trans I 103

549. Hartwig W, Born L (1987) Diastereoselective and Enantioselective Total Synthesis of the Hepatoprotective Agent Clausenamide. J Org Chem **52**: 4352
550. Fang XP, Anderson JE, Chang CJ, McLaughlin JL, Fanwick PE (1991) Two New Styryl Lactones, 9-Deoxygoniopypyrone and 7-*epi*-Goniofufurone, from *Goniothalamus giganteus*. J Nat Prod **54**: 1034
551. Fang XP, Anderson JE, Chang CJ, McLaughlin JL (1991) Three New Bioactive Styryllactones from *Goniothalamus giganteus* (Annonaceae). Tetrahedron **47**: 9751
552. Aoki M, Seebach D (2001) Preparation of TADOOH, a Hydroperoxide from TADDOL, and Use in Highly Enantioface- and Enantiomer-Differentiating Oxidations. Helv Chim Acta **84**: 187
553. Seebach D, Beck AK, Heckel A (2001) TADDOLs, Their Derivatives, and TADDOL Analogues: Versatile Chiral Auxiliaries. Angew Chem Int Ed **40**: 92
554. Sun BF, Hong R, Kang YB, Deng L (2009) Asymmetric Total Synthesis of (–)-Plicatic Acid *via* a Highly Enantioselective and Diastereoselective Nucleophilic Epoxidation of Acyclic Trisubstitued Olefins. J Am Chem Soc **131**: 10384
555. Gardner JAF, Barton GM, Maclean H (1959) The Polyoxyphenols of Western Red Cedar (*Thuja plicata* Donn.): 1. Isolation and Preliminary Characterization of Plicatic Acid. Can J Chem **37**: 1703
556. Vedal S, Chanyeung M, Enarson DA, Chan H, Dorken E, Tse KS (1986) Plicatic Acid Specific Ige and Nonspecific Bronchial Hyperresponsiveness in Western Red-Cedar Workers. J Allergy Clin Immunol **78**: 1103
557. Frew A, Chang JH, Chan H, Quirce S, Noertjojo K, Keown P, Chan-Yeung M (1998) T-Lymphocyte Responses to Plicatic Acid Human Serum Albumin Conjugate in Occupational Asthma Caused by Western Red Cedar. J Allergy Clin Immunol **101**: 841
558. Weissman DN, Lewis DM (2000) Is Specific Antibody Determination Diagnostic for Asthma Attributable to Low-Molecular-Weight Agents? Occup Med **15**: 385
559. Yamakoshi H, Shibuya M, Tomizawa M, Osada Y, Kanoh N, Iwabuchi Y (2010) Total Synthesis and Determination of the Absolute Configuration of (–)-Idesolide. Org Lett **12**: 980
560. Kim SH, Sung SH, Choi SY, Chung YK, Kim J, Kim YC (2005) Idesolide: A New Spiro Compound from *Idesia polycarpa*. Org Lett **7**: 3275
561. Tomizawa M, Shibuya M, Iwabuchi Y (2009) Highly Enantioselective Organocatalytic Oxidative Kinetic Resolution of Secondary Alcohols Using Chirally Modified AZADOs. Org Lett **11**: 1829
562. Guizzetti S, Benaglia M (2010) Trichlorosilane-Mediated Stereoselective Reduction of C=N Bonds. Eur J Org Chem **2010**: 5529
563. Drew MD, Lawrence NJ, Watson W, Bowles SA (1997) The Asymmetric Reduction of Ketones Using Chiral Ammonium Fluoride Salts and Silanes. Tetrahedron Lett **38**: 5857
564. Wallbaum S, Martens J (1992) Asymmetric Syntheses with Chiral Oxazaborolidines. Tetrahedron Asymmetry **3**: 1475
565. Corey EJ, Helal CJ (1998) Reduction of Carbonyl Compounds with Chiral Oxazaborolidine Catalysts: A New Paradigm for Enantioselective Catalysis and a Powerful New Synthetic Method. Angew Chem Int Ed **37**: 1986
566. Corey EJ, Roberts BE (1997) Total Synthesis of Dysidiolide. J Am Chem Soc **119**: 12425
567. Sabes SF, Urbanek RA, Forsyth CJ (1998) Efficient Synthesis of Okadaic Acid. 2. Synthesis of the C1-C14 Domain and Completion of the Total Synthesis. J Am Chem Soc **120**: 2534
568. Gunasekera SP, McCarthy PJ, Kelly-Borges M, Lobkovsky E, Clardy J (1996) Dysidiolide: A Novel Protein Phosphatase Inhibitor from the Caribbean Sponge *Dysidea etheria* de Laubenfels. J Am Chem Soc **118**: 8759
569. Tachibana K, Scheuer PJ, Tsukitani Y, Kikuchi H, Vanengen D, Clardy J, Gopichand Y, Schmitz FJ (1981) Okadaic Acid, a Cytotoxic Polyether from Two Marine Sponges of the Genus *Halichondria*. J Am Chem Soc **103**: 2469

Author Index

M. Waser, *Asymmetric Organocatalysis in Natural Product Syntheses*,
Progress in the Chemistry of Organic Natural Products 96,
DOI 10.1007/978-3-7091-1163-5, © Springer-Verlag Wien 2012

H

Subject Index

M. Waser, *Asymmetric Organocatalysis in Natural Product Syntheses*, Progress in the Chemistry of Organic Natural Products 96, DOI 10.1007/978-3-7091-1163-5, © Springer-Verlag Wien 2012

FSC
www.fsc.org
MIX
Papier | Fördert gute Waldnutzung
FSC® C083411

Kolibri 360 GmbH
Ferdinand-Jühlke-Straße 7
99095 Erfurt, Deutschland
produktsicherheit@kolibri360.de